Chemistry, Emission Control, Radioactive Pollution and Indoor Air Quality

Chemistry, Emission Control, Radioactive Pollution and Indoor Air Quality

Editor

Henri Duff

scitus
academics

Chemistry, Emission Control, Radioactive Pollution and Indoor Air Quality
Edited by **Henri Duff**

ISBN: 978-1-68117-317-7
Library of Congress Control Number: 2016937629

Notice

Reasonable efforts have been made to publish reliable data and views articulated in the chapters are those of the individual contributors, and not necessarily those of the editors or publishers. Editors or publishers are not responsible for the accuracy of the information in the published chapters or consequences of their use. The publisher believes no responsibility for any damage or grievance to the persons or property arising out of the use of any materials, instructions, methods or thoughts in the book. The editors and the publisher have attempted to trace the copyright holders of all material reproduced in this publication and apologize to copyright holders if permission has not been obtained. If any copyright holder has not been acknowledged, please write to us so we may rectify.

Printed in United States of America on Acid Free Paper

Preface

The atmosphere is our most valuable resource. The balance between its use and its protection is therefore a top priority for our civilisation. While many of us believe that air pollution is a problem that the modern world has largely resolved, it still seems to have a significant influence on the global environment. Radioactive pollution refers to air, water and other radioactive materials 'physical pollution. Natural radiation is also referred to as background radiation. It involves cosmic rays and reaches the earth's surface from space. The radioactive elements such as radium, uranium, thorium, radon, potassium and carbon are included. It happens in rock, soil and water. The radiations produced by man include the extraction and refining of plutonium and thorium. Indoor air pollution in developing countries is a major global health hazard. Acceptable levels of air pollution have been transgressed in many countries with ambitious economic growth targets. Serious problems related to respiratory disease have been identified with indoor and outdoor pollution throughout the world. This book deals with a number of air pollution issues—chemistry of air pollution; controlling processes of air pollution; radioactive pollution, and indoor air quality issues.

Table of Contents

CHAPTER 1

Carbon Dioxide Capture and Air Quality

Joris Koornneef[1], Toon van Harmelen[1], Arjan van Horssen[1] and Andrea Ramirez[1]

[1] Ecofys Netherlands, Utrecht, Netherlands

1. INTRODUCTION

Carbon dioxide (CO_2) is one of the most important greenhouse gases (GHG). The most dominant source of anthropogenic CO_2 contributing to the rise in atmospheric concentration since the industrial revolution is the combustion of fossil fuels. These emissions are expected to result in global climate change with potentially severe consequences for ecosystems and mankind. In this context, these emissions should be restrained in order to mitigate climate change.

Carbon Capture and Storage (CCS) is a technological concept to reduce the atmospheric emissions of CO_2 that result from various industrial processes, in particular from the use of fossil fuels (mainly coal and natural gas) in power generation and from combustion and process related emissions in industrial sectors. The Intergovernmental Panel on Climate Change (IPCC) regards CCS as "an option in the portfolio of mitigation actions" to combat climate change (IPCC 2005).

However, the deployment of CO_2 capture at power plants and large industrial sources may influence local and transboundary air pollution, i.e. the emission of key atmospheric emissions such as SO_2, NO_X, NH_3, Volatile Organic Compounds (VOC), and Particulate Matter ($PM_{2.5}$ and PM_{10}). Both positive as negative impacts on overall air quality when applying CCS are being suggested in the literature. The scientific base supporting both viewpoints is rapidly advancing.

The potential interaction between CO_2 capture and air quality targets is crucial as countries are currently developing GHG mitigation action plans. External and unwanted trade-offs regarding air quality as well as co-benefits when implementing CCS should be known before rolling out this technology on a large scale.

The goal of this chapter is to provide an overview of the existing scientific base and provide insights into ongoing and needed scientific endeavours aimed at expanding the science base.

The chapter outline is as follows. We first discuss the basics of CO_2 capture, transport and storage in section 2. In section 3, we discuss the change in the direct emission profile of key atmospheric pollutants when equipping power plants with CO_2 capture. Section 4 expands on atmospheric emissions in the life cycle of CCS concepts. We provide insights in section 5 into how air quality policy and GHG reduction policy may interact in the Netherlands and the European Union. Section 6 focuses on atmospheric emissions from post-combustion CO_2 capture. We highlight in section 7 the most important findings and provide outlook on (required) research and development.

2. CARBON DIOXIDE CAPTURE, TRANSPORT AND STORAGE

2.1. Co_2 Capture

The first step of the CCS chain is the capture process. A major element of this process comprises the separation of CO_2 from a gas stream. This can be the separation from produced natural gas, which often contains acid gases such as H_2S and CO_2. It also can be separated during the production of ammonia and during refining processes in the hydrocarbon industry. There is considerable less experience with removing CO_2 from flue gases at atmospheric pressure. This entails flue gases from power plants as well as industrial plants producing, for instance, steel, cement or iron. These large point sources form the largest potential for applying CO_2 capture. There are four approaches to capture CO_2 from large point sources: 1) Post-combustion capture; 2) Pre-combustion capture; 3) Oxyfuel combustion capture; 4) Capture from industrial processes.

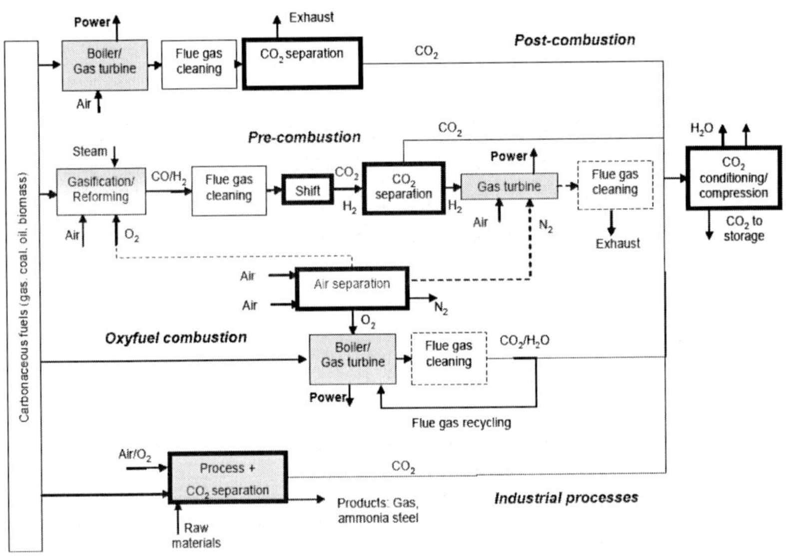

Figure 1. Simplified overview of the three CO_2 capture systems for power plants: post-, pre- and oxyfuel combustion. Grey components indicate power generation processes. Components with highlighted borders indicate processes causing a drop in generating efficiency. Components with dashed borders indicate optional processes. Note that natural gas reforming using steam is an endothermic process and therefore not a power generation process, hence the altered shading.

2.1.1. Post-Combustion Capture

CO_2 can be captured from the flue gas of a combustion process. This can be flue gas coming from any (pressurized) combustion in a boiler, gas turbine or industrial process yielding CO_2. Various capture mechanisms, or combinations of them, can be applied, being: phase separation, selective permeability and sorption. The last mechanism, sorption, is the most widely suggested mechanism to be used at large point sources. This mechanism encompasses chemical or physical absorption and also adsorption. In the CO_2 capture processes based on this mechanism a sorption medium, or a sorbent, is used. When these sorbents are in solution they are called solvents. The current research, development and demonstration (RD&D) focus is on using chemical and physical solvents to separate the CO_2 from the gas stream. Retrofitting existing power plants with CO_2 capture will highly likely be done with a chemical absorption based post-combustion capture technology.

The RD&D focus in post-combustion capture is mainly aimed at reducing energy requirement and capital cost trough developing and adapting solvents, optimizing the required process installations and integrating the capture system with the power generation process. The application of the capture process on contaminated flue gases, e.g. flue gases from coal fired power plants, is already commercially applied (Strazisar, Anderson et al. 2003). However, large-scale CO_2 capture as well as dealing with the contaminants in the flue gas remains a challenge.

2.1.2. Pre-Combustion Capture

Pre-combustion capture comprises a group of technologies that removes CO_2 before the combustion of the fuel. This requires a carbonaceous fuel to be broken down into hydrogen (H_2) and carbon monoxide (CO), i.e. syngas. To make CO_2 capture with high efficiencies possible, the syngas that is formed after steam reforming or partial oxidation/gasification has to be shifted after it is cleaned. The 'shift reaction', or 'water gas shift' (WGS) reaction, yields heat and a gas stream with high CO_2 and H_2concentrations. The CO_2 can then be removed with chemical and physical solvents, adsorbents and membranes.

For the near-term it is expected that chemical or physical solvents (or a combination) are used for the CO_2 removal. The CO_2 removal step yields relative pure CO_2 and a gas stream with a high hydrogen and low carbon content. The latter can be used for power production in for example a (modified) gas turbine. The gas with reduced carbon content can (after further purification) also be used in the production of synfuels, the refining of hydrocarbons or for the production of chemicals. (IPCC 2005)

For solid and liquid fuels, pre-combustion CO_2 capture can be applied in an IGCC (Integrated Gasification Combined Cycle) power plant. For gas fired power generation with pre-combustion capture other concepts are being studied (Ertesvag, Kvamsdal et al. 2005; Kvamsdal and Mejdell 2005;IEA GHG 2006c; Kvamsdal, Jordal et al. 2007).

The technology to capture CO_2 from the syngas generated in a gasifier can be considered proven technology, is commercially available and used for several decades in other applications than for electricity production. Examples are hydrogen, ammonia and synthetic fuel production (Nexant Inc. 2006). Also, reforming and partial oxidation of (natural) gas are already widely applied, e.g. for the production of hydrogen in the ammonia production process.

The pre-combustion concept has not yet been proven in an IGCC power plant. Proving its reliability and effectiveness in power plant concepts is therefore one of the main RD&D targets. In addition, improving the efficiency of the WGS step and integration of this process with CO_2 capture is also an area of research.

2.1.3. Oxyfuel Combustion

Oxyfuel combustion is based on denitrification of the combustion medium. The nitrogen is removed from the air through a cryogenic air separation unit (ASU) or with the use of membranes. Combustion thus takes place with nearly pure oxygen. The final result is a flue gas containing mainly CO_2 and water. The CO_2 is purified by removing water and impurities. The production of oxygen requires a significant amount of energy, which results in a reduction of the efficiency of the power plant. Further, the purification and the compression of the CO_2 stream also require energy.

The combustion with oxygen is currently applied in the glass and metallurgical industry (Buhre, Elliott et al. 2005; IPCC 2005; M. Anheden, Jinying Yan et al. 2005). Oxyfuel combustion for steam and power production using solid fuels has been at present only proven in test and pilot facilities. Oxyfuel combustion can also be applied in natural gas fired concepts. Power cycles for gaseous and solid fuels, however, vary significantly.

Although there are no significant differences compared to air firing of solid fuels, the combustion process and optimal configuration of the burners are considered to be the most important hurdles to overcome. In addition, the design and configuration of the flue gas cleaning section and CO_2purification section are challenges for the short-term. For the gas fired concepts, system integration and development of critical components hinder direct application on a commercial scale. Examples of critical components are the turbines and combustors for the near- and medium-term options and, additionally, the fuel reactors for the concepts in the longer term.

2.1.4. Capture From Industrial Processes

This group of technologies is often mentioned as the early opportunity for CCS at relative low cost. The total reduction potential due to

CO_2 capture from these point sources is however considered to be rather limited. Examples for industrial processes are: the production of cement, iron and steel, ethylene (oxide), ammonia and hydrogen. In addition, CO_2 can be captured from natural gas sweetening processes and from refineries (IPCC 2005). The capture processes applied are in general the same technologies as already described above.

2.1.5. Increased Primary Energy Use

When applying CO_2 capture, energy is needed to separate the CO_2 and compress the CO_2 to pressures required for transport. This energy consumption results in a reduction of the overall efficiency of for instance a power plant. This reduction is called the efficiency penalty, or energy penalty. Table 1shows typical energy penalties for power generation concepts with CO_2 capture.

Post-combustion CO_2 capture and capture using oxyfuel combustion of solid fuels show about equal increases in primary energy use. For post-combustion this increase is mainly determined by the heat requirement in the capture process. In oxyfuel combustion the separation of oxygen from the air is the main factor causing a drop in efficiency, i.e. about half of the efficiency penalty when considering a coal fired power plant (Andersson and Johnsson 2006). Both systems require significant compressor power to boost the CO_2 from atmospheric to transport pressures (i.e. > 100 bar). This compressor power is substantially lower in the pre-combustion technology as the CO_2 is removed under pressures higher than atmospheric. The required steam and the removal of chemical energy from the syngas in the process prior to CO_2 removal, the water gas shift reaction, contributes the most to the increase in primary energy use. The CO_2 removal itself requires less energy in this concept. Overall, the relative increase in primary energy is the lowest for the pre-combustion capture concepts. For the gaseous fuel fired concepts, the increase in primary energy requirement is relatively lower because of the lower carbon content per unit of primary energy.

Table 1. Simplified overview of energy conversion and CO_2 capture efficiencies of power plants equipped with various CO_2 capture technologies, after (Damen, Troost et al. 2006; Hetland and Christensen 2008)[a]PC = Pulverized Coal, NGCC= Natural Gas Combined Cycle, PFBC = Pressurized Fluidized Bed Combustion, GC = Gas Cycle, IGCC = Integrated Gasification Combined Cycle. [b]Efficiencies are reported based on the Lower Heating Value (LHV) and assuming a CO_2 product pressure of 110 bar.

Capture process	Conversion technology[a]	Generating efficiency[b] (%)	Energy penalty of CO_2 capture (% pts.)	Capture efficiency (%)
Post-combustion (chemical absorption)	PC	30-40	8-13	85-90
	NGCC	43-55	5-12	85-90
Oxyfuel	PC	33-36	9-12	90-~100
	GC and NGCC	39-62	2-19	50-~100
Pre-combustion	IGCC	32-44	5-9	85-90
	GC	43-53	5-13	85-~100

2.2. Co_2 Transport

The captured CO_2 can be transported as a solid, gas, liquid and supercritical fluid. The desired phase depends on whether the CO_2 is transported by pipeline, ship, train or truck. Of these options, transport by pipeline is considered the most cost-effective one. The transport of CO_2 by pipeline in the gas phase is not favourable for projects that require the transport of significant amounts of CO_2 over considerable distances. The disadvantageous economics (large pipeline diameter) and relative high energy requirement (due to the large pressure drop) are the reasons for this (IPCC 2005; Zhang, Wang et al. 2006). Increasing the density of CO_2 by compression renders the possibility to transport the CO_2 with less infrastructural requirements and lower cost.

There is worldwide experience in transporting CO_2 using the transport media mentioned above in the oil industry for enhanced oil recovery (EOR) by injecting CO_2 into an oil field. CO_2 transport by ship is being conducted on a small scale, but is being researched as a possibility to reach offshore storage capacity or as a temporary substitute for pipelines (IEA GHG 2004; Aspelund, Molnvik et al. 2006). Transport by ship can be economically favourable when large quantities have to be transported over long distances (>1000 km) (IPCC 2005). It requires the compression and liquefaction of the CO_2.

2.3. Co_2 Storage In Geological Formations

The last step in the CCS chain is the injection of CO_2 into geological formations. Alternatives to injection in geological formations are injection into the deep ocean and sequestration through mineral carbonation, but the current research focus is on storage in geological formations. CO_2 storage in these geological formations encompasses the injection of CO_2 into porous rocks that may hold or have held gas and or liquids. In literature, several storage media are proposed, especially: deep saline formations (aquifers); (near) empty oil reservoirs, possibly with enhanced oil recovery (EOR); (near) empty gas reservoirs, possibly with enhanced gas recovery (EGR) and deep unminable coal seams combined with enhanced coal bed methane production (ECBM). (Van Bergen, Pagnier et al. 2003; IPCC 2005)

The total CO_2 storage capacity ranges between 2 and 11 Tt. It should be stressed that high uncertainties still persist regarding the estimation of storage capacity due to the use of incomplete data or simplified assumptions on geological settings, rock characteristics, and reservoir performance (Bradshaw, Bachu et al. 2006). Despite the uncertainty of these estimates, the figures suggest that there is enough storage potential to support CO_2 emissions reduction with CCS for considerable time. In practice, matching the temporal and geographical availability of sources and sinks may become a bottleneck.

3. CHANGE IN KEY ATMOSPHERIC EMISSIONS DUE TO CO_2 CAPTURE

Key direct atmospheric emissions of specific interest for biomass and coal fired concepts are CO_2, NO_x, NH_3, SO_2, HCl, HF, VOC, PM, Hg, Cd, and other heavy metals. For gas fired concepts CO_2 and NO_x are the most dominant atmospheric emissions. Equipping power plants with CO_2 capture technologies affects both the formation and fate of many of these emissions. We limited our study to three main capture systems for the removal of CO_2 depicted in Fig. 1: post-combustion, pre-combustion and oxyfuel combustion.

The chemical absorption technologies that we reviewed in detail include technologies using alkanolamines, such as monoethanolamine (MEA), Fluor's Econamine FG+ and MHI's KS-1 solvent. Other technologies reviewed are based on absorption using chilled ammonia (NH_3), alkali salts (i.e. potassium carbonate -K_2CO_3) and amino salts. The post-combustion system can be applied to various energy conversion technologies. In this study we focus on its application to Pulverized Coal (PC), Natural Gas Combined Cycle (NGCC) and Pressurized Fluidized Bed Combustion (PFBC) power plants. The energy conversion technology that is envisaged using pre-combustion that is mainly investigated in this study is the Integrated Gasification Combined Cycle (IGCC) power plant. The energy conversion technologies using oxyfuel combustion that have been reviewed in this study more extensively are rather conventional PC and NGCC power plants. Advanced technologies briefly touched here include, for instance, chemical looping combustion.

A summary of emission factors for key atmospheric emissions reported in literature for these technologies is presented in Fig. 2. The main effects of CO_2 capture on atmospheric emissions are summarized below for the key atmospheric emissions.

3.1. Carbon Dioxide

CO_2 emissions predominantly depend on the type of fuel, on the efficiency of the energy conversion and of the removal efficiency of CO_2. The removal efficiency for the oxyfuel combustion concept is found to be the highest on average (95-98%), yielding the lowest CO_2 emissions for the gas fired conversion technologies (0-60 g/kWh). Post- and pre-combustion show about equal removal efficiencies of 87-90% and 89-95%, respectively. The typically higher conversion efficiency for gasification or reforming results however in typically lower net CO_2 emissions for the pre-combustion concepts (21-97 g/kWh) compared to the post-combustion concepts (55-143 g/kWh).

3.2. Sulphur Dioxide

In the coal fired power plants equipped with *post-combustion* CO_2 capture, SO_2 emissions are reduced significantly compared to a power plant without capture. One reason is that power plants with CO_2capture should be equipped with improved flue gas desulphurization (FGD) facilities (Tzimas, Mercier et al. 2007). Furthermore, additional removal in the post-combustion capture process is expected.Koornneef et al (2010) summarized reported values in literature and show that the minimum expected additional reduction per $MJ_{primary}$ compared to a power plant without CO_2 capture is approximately 40%; on average it is 85%.

For the amine based concept it is required to reduce the concentration of SO_x in the inlet gas of the CO_2 capture facility as these compounds may react with the solvent, which leads to the formation of salts and solvent loss. Knudsen et al. (2006; 2008) for instance reported a 40-85% uptake of total sulphur depending on the type of solvent[1] - used. Iijima et al. (2007) and Kishimoto et al. (2008) report that a minimum of 98% of the SO_2 is additionally removed[2] - before entering the CO_2 capture process. They state that then 'almost all' of the still remaining SO_2 is removed from the flue gas as salts. In literature studies additional SO_2 reductions of 90-99.5% are assumed (Rao and Rubin 2002;IEA GHG 2006a; Tzimas, Mercier et al. 2007; Koornneef, van Keulen et al. 2008).

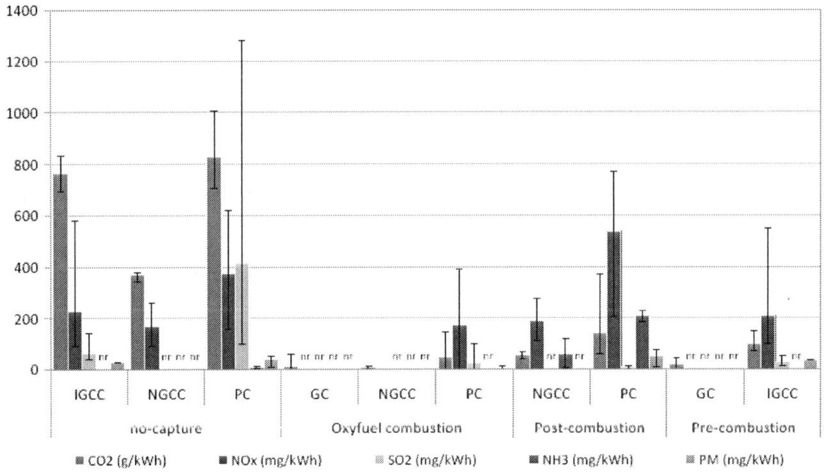

Figure 2. Atmospheric emissions of substances CO_2, NO_x, SO_2, NH_3 and particulate matter for various conversion technologies with and without CO_2 capture, adapted from (Koornneef, Ramirez et al. 2010). Ranges indicate maximum and minimum values reported. Note that emissions are based on various fuel specifications and on the configuration and performance of the power plant and CO_2capture process. 'nr' = 'not reported'.

The other post-combustion technology considered here uses chilled ammonia as solvent to remove the CO_2 from the flue gas. Remaining SO_2 in the flue gas can according to Yeh and Bai (1999) react with the ammonia solution to form the recoverable ammonium sulphate. All in all, it is expected that most of the acid gases can be removed from the flue gas when a proper design of the scrubbing process is applied (Yeh and Bai 1999). However, at present no quantitative estimates for additional SO_2reduction in the CO_2 absorption process based on chilled ammonia are available.

For *oxyfuel combustion* technologies the SO_2 emissions will generally decrease compared to conventional coal fired power plants. The reduction can be the result of several mechanisms: increased ash retention, enhanced efficiency of conventional FGD, co-injection and the possibility for new SO_xremoval technologies.

According to Buhre et al. (2005) and Anheden et al. (2005) the amount of SO_x formation per tonne of coal combusted is essentially unchanged when applying oxyfuel combustion. However, the composition and concentration of SO_x, constituting SO_2 and SO_3, does change as the flue gas stream is reduced in both volume and mass. A higher SO_x concentration in the flue gas may pose equipment corrosion problems. A possible positive effect is that it also may enhance the capture efficiency of the electrostatic precipitator (ESP) (Tan, Croiset et al. 2006). Another expected positive side effect is that a higher SO_x concentration may increase the removal efficiency[3] - of FGD technologies. Moreover, the reduced flue gas stream allows for smaller equipment. (Marin and Carty 2002; Chatel-Pelage, Marin et al. 2003; Chen, Liu et al. 2007; WRI 2007)

The issues, challenges and design considerations taken into account when designing the flue gas cleaning section for *oxyfuel* combustion are presented in (Yan, Anheden et al. 2006). There, possible configurations for flue gas cleaning are predominantly based on (adapted) conventional flue gas cleaning technologies. The additions compared to a conventional configuration consisting of an SCR, ESP and FGD, are a flue gas cooler (FGC) and CO_2 compression & purification process. The FGC is aimed to reduce the temperature, acidic substances (SO_2 between 93 and 97%, SO_3 between 58 and 78%), water content (>85%) and particulates (>90%) in the flue gas prior to compression. In the following compression & purification step additionally NO_x, SO_x, HCl, water and heavy metals are removed as condensate from the compressors, and with the use of an activated carbon filter and an adsorber (Burchhardt 2009; Thébault, Yan et al. 2009; Yan, Faber et al.

2009). Overall, a deep reduction of SO_2 and NO_x emissions is expected to be possible with oxyfuel combustion, although R&D is required to better understand the behaviour of these substances in the CO_2 compression & purification process.

Co-injection of sulphur compounds into the underground together with the CO_2 is technically possible. Another possibility is the removal of sulphur compounds in condensate streams after compression of the flue gas. Both options would make the FGD section redundant. As suggested by White et al. (2008) the SO_2 may be recovered from the CO_2 stream in the form of sulphuric acid (H_2SO_4) through reaction with NO_2. Experiments indicate SO_2 conversion efficiencies between 64 and ~100% depending on process conditions (White, Torrente-Murciano et al. 2008).

In circulating fluidized bed boilers often limestone is injected into the furnace to control SO_xemissions. In the case of oxygen firing the in-furnace desulphurization efficiency with limestone is found to be between 4 and 6 times higher compared to air firing (Buhre, Elliott et al. 2005; ZEP 2006).

The variance shown in Fig. 2 is due to parameters that may vary case by case, e.g. the sulphur content in the coal, uncontrolled SO_x emission (including ash retention), removal efficiency of the FGD section, removal in CO_2 purification section and the degree of co-injection.

IGCC power plants have low SO_2 emissions, either with or without *pre-combustion* CO_2 capture. This is due to the high (typically > 99%) removal efficiencies of sulphur compounds (H_2S and COS) in the acid gas removal section and adjoined facilities. The application of pre-combustion CO_2 capture in an IGCC is assumed to enhance the SO_2 removal. The application of CO_2 capture is likely to result in a decrease of the emission of SO_2 per $MJ_{primary}$, but depending on the efficiency penalty may result in an increase per kWh. Both increase and decrease per kWh have been reported in literature. The reduction per $MJ_{primary}$ is expected to be lower compared to the post-combustion and oxyfuel combustion technologies (see Fig. 2). With pre-combustion it is also possible to yield a stream of CO_2 with H_2S and co-inject this into the underground. This may however complicate the transport and storage process. Also, it may be prohibited by national law and varies per country.

3.3. Nitrogen Oxides

If an amine based solvent is used for *post-combustion* capture, the reduction of NO_x emissions per $MJ_{primary}$ is expected to be small, i.e. between 0.8 and 3%[4] - (Knudsen, Vilhelmsen et al. 2006;Kishimoto, Hirata et al. 2008). CO_2 capture requires a significant increase in primary energy use resulting in a net increase in NO_x emissions per kWh. For the chilled ammonia technology, the NO_xemissions are not known to be affected by the CO_2 absorption process. It is, therefore, likely that emissions will increase proportionally with the increase in primary energy use.

For *oxyfuel* combustion, in general, net NO_x emissions per $MJ_{primary}$ are likely to decrease compared to conventional coal fired power plants. The two most important factors are that coal fired oxyfuel power plants are likely to show lower levels of NO_x formation in the combustion process and that further high degree of removal of NO_x in the CO_2 treatment train is possible.

NO_x emission reduction and underlying mechanisms are fairly well understood for the oxyfuel combustion technology. NO_x formation during oxyfuel combustion is found to be lower as thermal NO_xformation is suppressed and fuel NO_x is reduced (Croiset and Thambimuthu 2001; Buhre, Elliott et al. 2005; Tan, Croiset et al. 2006; WRI 2007). Overall, the reduction potential for NO_x formation of oxyfuel combustion is according to several experiments in the range of 60-76% (Chatel-Pelage, Marin et al. 2003; Buhre, Elliott et al. 2005; Farzan, Vecci et al. 2005; Andersson 2007; Yamada 2007). However, also no reduction has been found in some experiments (Anheden, Jinying Yan et al. 2005).

The final emission of NO_x depends also on the flue gas treatment section. The flue gas has a high CO_2concentration, but also contains NO_x, Ar, N_2, O_2 and SO_2 when it enters the CO_2 treatment train. There are several options for the treatment of the raw CO_2 stream. None of them requires a $DeNO_x$facility like SCR or SNCR[5] - (DOE and NETL 2007). The first option is to co-inject the NO_x together with the CO_2. This requires only compression and drying of the flue gas stream. The second option is to purify the CO_2 with multiple auto-refrigeration flash steps. The gaseous pollutants are, in that case, separated from the CO_2 stream to a high degree and vented into the atmosphere. The remaining fraction is co-injected. A $DeNO_x$ installation may be used to clean the vent stream (IEA GHG 2006b). Another concept is suggested and tested by White et al. (2006; 2008) and incorporates compression of the flue gas and removal of NO_x in the form of nitric acid (HNO_3) through a series of reactions[6] - . Preliminary results suggest that 48-90% of the NO_x is converted to nitric acid[7] - and can consequently be removed from the CO_2 stream.

The oxyfuel combustion variant shows no NO_x emissions from gas fired power plants equipped with CO_2 capture. This estimate is based on one literature source only, i.e. see (Davison 2007). This may result in an underestimation of NOx emissions. As the purity of the oxygen stream is in practice not 100%, some nitrogen may still be present in the combustion air, causing some NO_x formation (IEA GHG 2006c). Whether this is co-injected or separated depends on process configuration.

During normal operation of the IGCC with *pre-combustion* CO_2 capture, NO_x will be mainly formed during the combustion of the hydrogen rich gas with air in the gas turbine. The application of CO_2capture in an IGCC will decrease the NO_x emissions per $MJ_{primary}$ as relatively less gas is combusted in the gas turbine per unit of primary energy input. This outcome strongly depends on the assumption that the issue of NO_x formation in a gas turbine fired with fuel gas with a high hydrogen content is solved by turbine manufacturers. The flame temperature is namely dependent on the gas composition and heating value. Both of these will change when applying CO_2 capture. If dilution with steam or nitrogen is not applied, the flame temperature during firing of hydrogen rich fuel will increase resulting in an increase in NO_x formation. Consequently, emissions per kWh can also become higher when applying CO_2 capture. The uncertainty is thus higher than the range indicated in Fig. 2. This is however not quantified. (Chiesa, Lozza et al. 2005; IEA GHG 2006c; Davison 2007; DOE/NETL 2007; Tzimas, Mercier et al. 2007)

For gas fired concepts equipped with pre-combustion capture, NO_x emissions are uncertain but expected to be typically higher than for conventional state-of-the-art NGCC cycles (Kvamsdal and Mejdell 2005).

Further NO_x emission reduction can be achieved by adding an SCR process. A possible trade-off for SCR application is the emission of unreacted ammonia, or ammonia slip. This is especially the case when the SCR is applied on exhaust gases with low NO_x concentrations. Ammonia slip from a SCR are however very small (<5 ppmv) and is assumed comparable to normal air combustion in a pulverized coal power plant and a NGCC power plant. An optimum between NO_x reduction and ammonia slip is however to be determined (Rao 2006).

3.4. Ammonia

Ammonia slip from DeNOx facilities is the main source of NH_3 emissions from conventional fossil fuel fired power plants without CCS.

A significant increase of NH_3 emissions may be caused by oxidative degradation of amine based solvents that possibly will be used in *post-combustion* CO_2 capture. In the chilled ammonia technology, the unwanted emission of NH_3 from the CO_2 capture process is a serious challenge (Yeh and Bai 1999). This emission is expected to be significantly reduced by adding a water wash section at the outlet of the CO_2 capture process and by adaptations in the capture process (Yeh and Bai 1999;Corti and Lombardi 2004; Kozak, Petig et al. 2008).

As indicated, the uncertainty regarding the estimation of NH_3 emissions can be considered high as the scientific literature reports a variety of values (Rao and Rubin 2002; IEA GHG 2006a;Knudsen, Jensen et al. 2008). Furthermore, new solvents and additional treatment options are possible to prevent or mitigate the emission of ammonia. The ranges shown in Fig. 2 are thus rather conservative estimates.

For *oxyfuel* combustion, no quantitative estimates for ammonia emissions are known to be reported.

Ammonia formed during gasification is effectively removed in the gas cleaning section in an IGCC with *pre-combustion*. Therefore, emissions are considered negligible.

3.5. Volatile Organic Compounds

No quantitative estimates for VOC emissions could be derived due to a lack of quantitative information in the pertaining literature.

It is possible that VOC emissions are not significantly influenced by the *post-combustion* CO_2 capture process. In that case the VOC emissions will increase with the increase in primary energy use. However, degradation of amine based solvents may result in the emission of volatile substances, e.g. formaldehyde, acetone, acetaldehyde (Knudsen, Jensen et al. 2008). New solvents are being developed and tested that do not show these degradation products (Hopman. 2008; Knuutila, Svendsen et al. 2009).

No clear information was found on the effect of *oxyfuel* combustion on the formation, reduction and final emission of VOC. However, the oxygen rich conditions during combustion may have an effect on VOC formation. The fate of the formed VOC is uncertain, but it is plausible that a part of the VOCs is either co-injected or vented from the CO_2 purification section (Harmelen, Koornneef et al. 2008).

In IGCC power plants there are two main origins of VOC emissions: the gas turbine section and the fuel treatment section. The formation of VOC in the first is expected to be reduced due to *pre-combustion* CO_2 capture and the associated higher hydrogen content of the fuel gas. Quantitative estimates for the reduction of VOC are however not available. The emissions from the fuel treatment section are expected to remain equal per $MJ_{primary}$. VOC emission reporting for an IGCC operating in the Netherlands does not provide decisive insights into which section is the dominant source of VOC (NUON 2005; NUON 2006). The net effect of both may thus be an increase or decrease per kWh. For gas fired cycles, the replacement of natural gas with hydrogen rich fuel gas is expected to lower the emission of VOC.

3.6. Particulate Matter

Often no distinction is made in the consulted literature between various sizes[8] - of emitted particulate matter in emission reporting. In this review, therefore also no distinction could be made between size fractions.

The high variance for *post-combustion* capture technologies for solid fuel fired power plants stands out in Fig. 2. The variance represents the varying assumptions in literature. On the one hand, some scholars assume a deep reduction of PM due to the application of post-combustion CO_2 capture; on the other hand, other scientists assume that it will not have an effect on PM emissions. Results from an amine based post–combustion capture demonstration project however indicate a decrease in emission of particulate matter of 64-80%[9] - per $MJ_{primary}$ (Kishimoto, Hirata et al. 2008). Also Kozak et al. (2008) suggest a decrease with the use of chilled ammonia technology[10] -. An increase in emission per $MJ_{primary}$ is never assumed. Together with the energy penalty due to CO_2 capture, PM emissions may however increase per kWh.

The low particulate matter emissions found for the *oxyfuel* combustion technology are partly due to the enhanced removal efficiency of the ESP[11] - that is possible during oxyfuel combustion. Particulates may also be partially co-injected with the CO_2 stream. Another possibility is that particulates are vented from the CO_2 treatment section. Yet another option is that PM is removed with the condensate stream that is formed when SO_2 and NO_x are removed as sulphuric and nitric acid, as mentioned earlier. All together, PM emissions are estimated to be very low.

IGCC power plants are assumed to have lower PM emission factors compared to other conversion technologies and types of power plants. *Pre-combustion* CO_2 capture has virtually no influence on the emission of PM (per $MJ_{primary}$) from an IGCC.

Although no quantitative estimates are available, it may be possible that PM emissions, in specific $PM_{2.5}$ emissions, will be lower due to the enhanced capture of sulphur compounds from the syngas, which is expected to reduce the formation of sulphates, which are characterized as $PM_{2.5}$.

3.7. Other Atmospheric Emissions Of Interest

Fig. 1 shows that the *post-combustion* CO_2 capture process is situated after the flue gas cleaning section. Depending on the type of solvent that is used, impurities need to be removed from the flue gas in order to limit operational problems. When MEA is used, its consumption in the capture process is mainly caused by degradation by oxygen and impurities in the flue gas. Important impurities are sulphur oxides (SO_x), nitrogen dioxide (NO_2), hydrogen chloride (HCl), hydrogen fluoride (HF) and particulate matter as they react with the MEA or cause foaming of the solvent. This may result in reduction of HF and HCl emissions. Estimates in literature vary but are as high as 90-95%.

Power plants equipped with CO_2 capture should thus be equipped with highly efficient flue gas desulphurization (FGD), $DeNO_x$ installations and electrostatic precipitators (ESP) and/or fabric filters to remove PM. Also, the flue gas typically requires cooling before it is processed in the CO_2 capture installation. In the CO_2 capture process also some of these substances are partially removed. The capture process is thus expected to affect (i.e. lower) the emission of these air pollutants directly and indirectly.

The consumption of solvent in the capture process is an important driver for solvent development as solvent loss deteriorates operational economics and has environmental consequences. The consumption of the solvent varies per type of solvent but is for post-combustion typically in the order of 1-2 kg/tonne CO_2 captured. Recent pilot plant test campaigns report solvent consumption rates for MEA at 0.3 kg/tonne captured (Moser et al. 2011). Typically, the consumption of MEA is higher compared to its alternatives. Moreover, the consumption of solvent used in IGCC with or without pre-combustion concepts can be considered very low, although an increase is expected when CO_2 capture is applied.

The higher oxygen concentration in the flue gas from natural gas combustion possibly results in higher oxidative degradation of solvents. MEA is to instance susceptible for this type of degradation (Supap, Idem et al. 2009). However, as other impurities such as SO_2 and PM are virtually not present in the flue gas, overall degradation and consumption is considerably lower compared to coal fired power plants.

For some post-combustion variants additional atmospheric emissions are expected. More specifically, the emission of solvent or degradation products of the solvent are currently of high interest (see section 6). For MEA based solvents this may be direct MEA emissions. The exact quantity of this 'MEA slip' (estimates range between 1 and 4 ppmv) and possible effects on the environment, including human safety, are not fully known but are intensively researched. In addition, solvent additives (e.g. corrosion inhibitors) may result in trace emissions of heavy metals (Thitakamol, Veawab et al. 2007).

For the chilled ammonia process solvent emissions may be NH_3 (Yeh and Bai 1999; Corti and Lombardi 2004; Kozak, Petig et al. 2008). The alkanolamine based solvents may result in the emission of VOC and NH_3 due to the degradation of the solvent (Strazisar, Anderson et al. 2003; Rao, Rubin et al. 2004; Knudsen, Jensen et al. 2008). Korre et al. (Korre, Nie et al. in press) report that NH_3 emission from using the MHI KS-1 solvent is expected to be higher than from using MEA or potassium carbonate[12] -. Contrarily, IEA GHG (IEA GHG 2006a) reports lower values for NH_3 emissions for the MHI KS-1 process compared to Fluor's process based on MEA.

For a K_2CO_3 based sorbent the slip into the atmosphere is considered negligible. Furthermore, this substance is considered to be less toxic to the environment (Oexmann and Kather 2009; Smith, Ghosh et al. 2009). K_2CO_3 may however require the addition of promoters to increase the reaction rate. Some promoters, like arsenic trioxide and piperazine, are known to be toxic (Smith, Ghosh et al. 2009).

Allaic and Jaspers (Allaie and Jaspers 2008) claim that the use of amino salts does not result in ammonia formation, losses due to evaporation and virtually nihil emissions of the solvent.

Mercury (Hg) and other heavy metals may be partially removed in the CO_2 capture process. Measurements on reclaimer bottoms have indicated that mercury is present in the bottoms (Strazisar, Anderson et al. 2003). A recent study however indicates that a combined process of removing mercury and CO_2 would not lead to significant reductions, i.e. below 15% (Cui, Aroonwilas et al. 2010).

In coal and biomass fired *pre-combustion* concepts using solvents, no solvent emission to the air is expected during normal operation as any slip of the solvent would be combusted in the gas turbine or end up in the CO_2 stream. For this technology, co-sequestration of H_2S is a technical possibility and co-injection is common in for instance Canada (Thomas and Benson 2005).

Carbon monoxide emissions from an IGCC with capture are reduced as the WGS process is introduced, converting CO and H_2O into H_2 and CO_2.

A second source of CO emissions is the pre-preparation of the fuel, i.e. storage and grinding. The latter emissions will not be directly affected by the CO_2 capture process, although per kWh those emissions will probably increase as more fuel has to be stored and handled per kWh.

No solvents are used in the *oxyfuel* combustion system. Additional gaseous emissions per primary energy input are thus not expected. In literature it is however suggested that in oxyfuel concepts, due to higher oxygen concentrations, a larger part of elemental mercury (Hg) is converted to ionized Hg species, which will possibly result in higher capture efficiencies of Hg in flue gas cleaning sections ($DeSO_x$ and DeHg). (Marin and Carty 2002; Chatel-Pelage, Marin et al. 2003) (WRI 2007) This may be an additional benefit of CO_2 capture with oxyfuel combustion.

For the advanced oxyfuel concepts, chemical looping combustion[13] - , the loss of metal oxides to the atmosphere may be a concern. Metal oxides are used as oxygen carriers to physically separate the reduction and oxidising step in this power cycle. These metal oxides may contribute to the environmental impacts of energy supply with this concept as it might bring forward direct environmental impacts, i.e. some metals are considered toxic. Also, these oxygen carriers may bring forward environmental impacts in their life cycle, e.g. during mining, treatment and disposal.

4. ATMOSPHERIC EMISSIONS ACROSS THE VALUE CHAIN OF CCS

Life Cycle Assessment (LCA) is today one of the most used tools for evaluating the potential environmental impact of products and materials. LCA is a technique for assessing the environmental aspects and potential impacts associated with inputs and outputs of a product system. In the case of CCS, a full LCA includes the production of the fuel carrier (e.g., mining of coal), fuel transport, power production, CO_2 capture, CO_2 transport and CO_2 storage (see Fig. 3). Note that for most studies found in the literature, including those on CCS, emissions from the infrastructure and the extraction of raw materials other than fuel tend to be excluded since they are assumed to be relatively small in comparison to primary burdens or there is lack of data that does not allow for a reliable analysis.

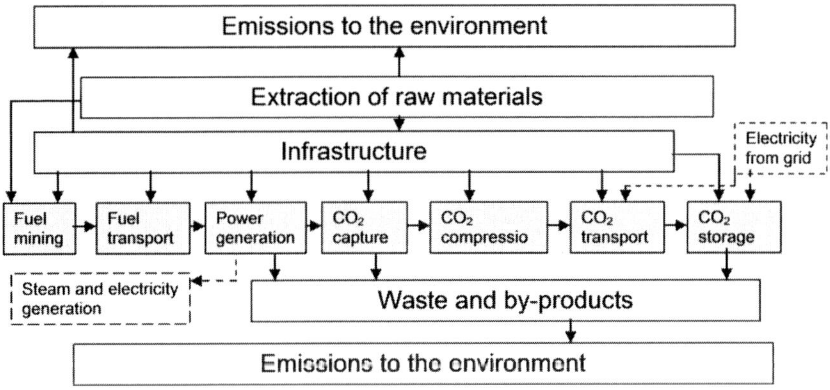

Figure 3. Schematic overview of step chains of fossil fuel production with CCS and interactions with other systems.

In this section, we consider the results found by 25 studies published between 1995 and 2009. Each reviewed study typically addresses different impact categories. The focus of this section is on the atmospheric emissions of CCS during its life cycle[14] - and, therefore, only the following categories will be examined in detail: CO_2 emissions, SOx/NOx and Particulate matter. The literature used to assess the emissions is listed in Table 2.

Table 2. Literature sources used to assess the emissions of the CCS value chain.

Capture technology type	Literature sources
PC with CCS	(Rao and Rubin 2002;Spath and Mann 2004;IEA GHG 2006a;Khoo and Tan 2006;Tzimas, Mercier et al. 2007;Viebahn, Nitsch et al. 2007;Weisser 2007;Koornneef, van Keulen et al. 2008;Odeh and Cockerill 2008;RECCS 2008;Bauer and Heck 2009;Markewitz, Schreiber et al. 2009;NEEDS 2009;Nie 2009;Pehnt and Henkel 2009;Schreiber, Zapp et al. 2009;Korre, Nie et al. 2010)
IGGC with CCS	(Waku, Tamura et al. 1995;Akai, Nomura et al. 1997;Lombardi 2003;Tzimas, Mercier et al. 2007;Viebahn, Nitsch et al. 2007;Weisser 2007;Odeh and Cockerill 2008;RECCS 2008;Pehnt and Henkel 2009)
Oxyfuel with CCS	(Viebahn, Nitsch et al. 2007;RECCS 2008;Bauer and Heck 2009;NEEDS 2009;Nie 2009;Pehnt and Henkel 2009)
Natural gas with CCS	(Lombardi 2003;Spath and Mann 2004;Sundkvist, Klang et al. 2004;IEA GHG 2006a;Hertwich, Aaberg et al. 2008;Odeh and Cockerill 2008;RECCS 2008;Bauer and Heck 2009;NEEDS 2009)

4.1. Co$_2$ Equivalent Emissions

The main goal of CCS is to reduce CO_2 emissions and consequently, Global Warming Potential (GWP). For pulverized coal-fired power plants with post-combustion capture technology using MEA a range in GWP over the life cycle of 79-275 gCO_2eq/kWh is reported (range for PC without CCS is in the range 690 to 1100 gCO_2eq/kWh). Where PCs without CCS have a share of power plant operation in life cycle GWP of about 80-95%, installing CO_2 capture decreases this share to about 43-60%. Thus, the deployment of CCS results in a pronounce increase in the share of indirect CO_2eq. emissions in the complete life cycle[15] - .

In the case of IGCCs with pre-combustion CO_2 capture, GWP values reported are in the range 110 to 181 gCO_2eq/kWh (the range for IGCCs without CCS is 666 to 870 gCO_2eq/kWh). Lignite-fired IGCCs with CCS have almost 20% less absolute emissions compared to hard coal-fired IGCCs with CCS. Installing CCS results in a reduction of about 82 to 87% for lignite-fired IGCCs with CCS relative to IGCCs without CCS, while for hard coal-fired IGCCs the relative differences are in the range of 69 to 81%.

Interestingly, hard coal-fired power plants with CCS technology are reported as having between 20% (IGCC with CCS) and 30% (PC with CCS) more GHG emissions than similar lignite-fired power plants with CCS, while without CCS technology the hard coal-fired power plants have about 10% lower emissions than lignite-fired power plants. This is due to a typically larger share of upstream emissions (e.g. fuel extraction and processing) for hard coal-fired power plants than for lignite-fired power plants. Lignite-fired power plants are often directly located at the mining site ('mine-mouth' operated) which results in lower transport emissions. As these upstream emissions are not reduced by CCS technology, but mainly increase due to the energy penalty, the GHG emissions from mine-mouth based power plants can be reduced further when implementing CCS.

The range in GWP for oxyfuel power plants with CCS is 25-176 gCO2eq/kWh. The relative decrease in GWP ranges from 78% to 97%. Specifically for hard coal-fired power plants, the range in the relative difference is smaller (78% to 87%).

For NGCCs with post-combustion using MEA, the GWP are in the range 75-245 gCO2eq/kWh, which is about 51-80% less than the values reported for NGCCs without CCS. The two studies (Spath and Mann 2004; Odeh and Cockerill 2008) reporting the lower value range (51-58%) also report about 25% indirect emissions for NGCCs without CCS technology compared to about 12-15% reported in other studies. Both studies state that in the case of NGCCs the amount of methane leakage from natural gas extraction and transport has a significant effect on life cycle GHG emissions and is more inopportune in the case with CCS due to the increase in primary energy consumption. It is, however, unclear if the other studies include this methane leakage in the reported values.

4.2. Particulate Matter

PM_{10} emissions reported for the life cycle of PC power plants with post-combustion using MEA range between 0.013 and 0.434 gPM_{10}/kWh while $PM_{2.5}$ for the same type of plants are reported between 0.05 to 0.07 gPM2.5/kWh. PC plants without CCS report PM_{10} in the range 0.009 to 0.35 gPM_{10}/kWh and $PM_{2.5}$ in the range 0.009 to 0.35 gPM_{10}/kWh. Contrary to the results found for GWP, no clear difference is reported for hard coal-fired and lignite-fired power plants.

Only two studies (Viebahn, Nitsch et al. 2007; RECCS 2008) report the contribution of the different part of the CCS chains. In these studies, the contribution of the PC plant with CO_2 capture is estimated at 33% and 45%, which is lower than the estimated contribution of a similar PC plant without CCS (60% and 65%, respectively).

The amount of studies reporting PM emissions for other CO_2 capture technologies is limited. One study by Odeh and Cockerill (2008) reports PM emissions for IGCCs with and without CCS technology (0.004 g/ kWh in both cases). The value is lower than those reported for PCs due to the high removal of PM in the syngas (to avoid detrimental effects in the turbine). NEEDS (2009) reports PM values for NGCCs equipped with MEA based post-combustion technology in the order of 0.005-0.006 gPM_{10}/kWh and 0.009-0.010 $gPM_{2.5}$/kWh (the values for a NGCC without CCS are in the range of 0.003-0.012 gPM_{10}/kWh and 0.007-0.008 $gPM_{2.5}$/kWh). In this case, PM emissions are mainly associated with NOx emissions (which are PM precursors) from the power plant and the winning of natural gas. Values for oxyfuel power plants with CO_2 capture are also reported by NEEDs (2009). For PM_{10} the range reported is 0.012 to 0.025 gPM_{10}/kWh while for $PM_{2.5}$ this is 0.07 to 0.36 $gPM_{2.5}$/kWh.

4.3. No_x And So_x

Table 3 shows the ranges found in the literature for the life cycle of power plants with and without CCS. The number of studies reporting specific results, particularly for IGCC and oxyfuel plants with CCS, is quite limited making it difficult to draw robust conclusions.

Results for PC with post-combustion capture using MEA indicate an increase in the amount of NOx during the life cycle. The partial removal of NOx during the capture process is not large enough to offset the increase in emissions caused by the additional fuel needed to compensate the energy penalty. Literature appears less clear on the impact of post-combustion on SOx levels, with some studies showing a decrease in emissions while others indicating an increase on emissions. Koornneef et al., (2008) indicates that a decrease in both NOx and SOx emissions associated with the transport of coal can be expected since stringent regulations are anticipated to reduce sulphur content in marine fuel and to limit NOx emissions during ship transport.

In the case of NGCC plants with post-combustion MEA, SOx emissions are reported to increase over the life cycle. The level however remains well below those of PC plants due to the very low sulphur content of natural gas.

Table 3. Emissions of SOx and NOx reported in the literature and relative change compared to a similar plant without CCS (N= number of studies).

Type of plant	NOx			SOx		
	With CCS (gNOx/kWh)	% change	N	With CCS (gSO_2/kWh)	% change	N
PC	0.58- 1.39	+13% to+49%	7	0.08-0.84	-60% to +20%	6
IGGC	0.10	-16%	1	0.33	+10	1
Oxyfuel	0.27-0.60	---	1	0.11-0.35	---	1
NGCC	0.13-0.30	-50% to +15%	5	0.01-0.16	+14 % to +100%	3

5. CASE STUDIES – NETHERLANDS AND EUROPE

In this section we discuss two case studies showing the consequences of implementing CCS in the power and heat sector on non-GHG emissions that affect air quality. We consider two geographical regions: the Netherlands and the European Union.

5.1. Scenario Study For The Dutch Power and Heat Sector

The BOLK research programme was initiated by the Dutch Ministry of Housing, Spatial Planning and the Environment (VROM) to acquire more detailed information on the synergy and/or trade-offs of GHG mitigation policies and transboundary air pollution (AP) policies (Harmelen, Koornneef et al. 2008; Horssen, Kuramochi et al. 2009). Part of that programme was aimed at assessing CCS technologies and includes scenario analyses for key atmospheric emissions (NO_x, SO_2, PM_{10} and NH_3) from the power sector. Three scenarios were developed and compared to the 2006 emission level (seeFig. 4):

1. Power sector emissions without CCS : No CCS is applied to the power plant sector in 2020
2. 2020 with CCS -S1: CCS is applied to two new coal fired power plants, 1 with post-combustion and 1 with pre-combustion.
3. 2020 with CCS -S2: CCS is applied to all new coal fired power plants, 3 with post-combustion and 1 with pre-combustion.

The impact of applying CO_2 capture technologies to the power generating sector is assessed in that study for several view years (i.e. up to 2050). For the view year 2020 the scenario is based on the actual and planned power plants.CO_2 emissions from the sector are assumed to increase from 38 in 2006 to 62 Mtonne in 2020, without the introduction of CCS. In 2020, up to 24 Mtonne CO_2 could be avoided when equipping new coal fired power plants with CO_2 capture.

The results of that scenario study are shown in Fig. 4. Emissions of NO_x, SO_2, PM_{10} and NH_3 in the sector are estimated to increase in the reference scenario for the year 2020, due to the increase in Dutch coal fired capacity without CCS. The introduction of CCS (only post- and pre-combustion) is expected to lead to a further increase of NO_x (up to 1.5 ktonne), PM (up to 70 tonne) and NH_3 emissions (up to 0.7 ktonne). SO_2 emissions decrease below the 2006 level. The introduction of CCS leads to a relative large increase (from 0.1 ktonne to up to 0.8 ktonne) in the overall low contribution of the power generation sector to the NH_3 emissions. For the scenario analyses, the emission data in relative old publications associated with the use of ethanolamines were used. Developments are still going on to reduce the solvent degradation and with it the emissions from solvents. With the improvement of the solvent technology, NH_3 emissions will be strongly reduced.

The additional cost of mitigating key atmospheric emissions are roughly assessed and compared to the reference scenario. The results showed that the cost mainly consist out of the increased costs for air pollution control needed to counteract the projected capacity increase of the power plant sector. The additional mitigation costs due to CCS are estimated to be small compared to those costs. The costs of NO_x and PM_{10} dominate the overall mitigation costs of approximately 50 million Euros per year.

The mitigation costs of NO_2, PM_{10} and NH_3 compared to the reference scenario are positive, because of the increase of the emissions in the CCS scenarios. Mitigation cost for SO_2 are negative due to the emission reduction of this substance in CCS equipped power plants.

5.2. Scenario Study For The European Power and Heat Sector In 2030

In a study by Koornneef et al (2010) the trade-offs and synergies between climate and air quality policy objectives for the European power and heat (P&H) sector were quantified. The analysis includes assessing the impact of applying CO_2 capture in the European P&H sector on the emission level of key air pollutants in 2030. A model was developed with the assumption that all power plants built between 2020 and 2030 are equipped with CO_2 capture and that all plants built between 2010 and 2020 are retrofitted[16] - with CO_2 capture before 2030. Four scenarios were investigated: one without CCS (baseline) and three with CCS. Each one focuses on a different CO_2 capture system (post-combustion, oxyfuel combustion and pre-combustion).

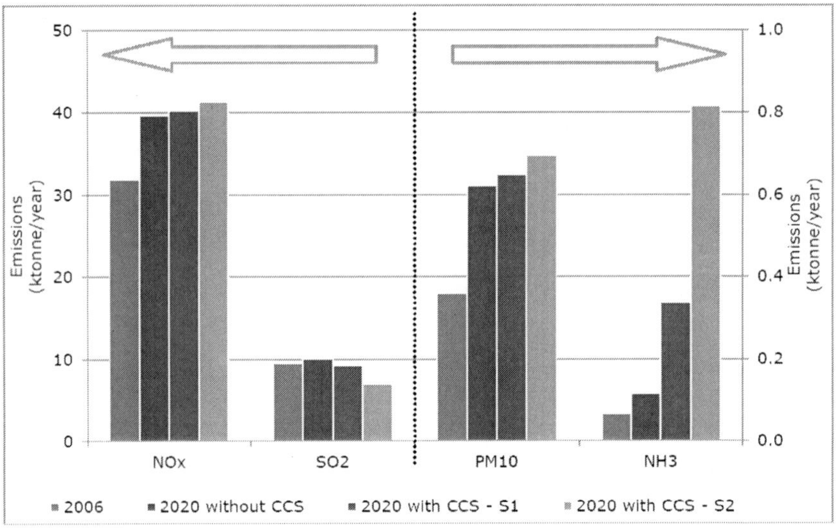

Figure 4. Emissions of NO_2, SO_2 (ktonne/year) on left axis and of PM_{10}, NH_3 (ktonne/yr) on right axis, in the Dutch power plant sector in 2020.

The first scenario without climate measures was drawn entirely from the GAINS[17] - model developed by the IIASA (IIASA 2008). The emission levels of NECD substances in 2030 were defined by combining sector activity and emission factors for P&H plants with and without CO_2 capture.

In the three scenarios with CO_2 capture the share of power plants equipped with CO_2 capture technology was determined by estimating for each country the additional sector activity (in primary energy use) in the baseline scenario per combination of conversion technology and fuel for the periods 2010-2020 and 2020-2030.

Results show a reduction in GHG emissions compared to the baseline scenario between 7% and 16% for scenarios with CCS penetration in the European power and heat sector. This comes with an increase in total primary energy use in the EU of approximately 1-5%. In the power and heat sector alone this increase is between 2% and 17%.

SO_2 emissions are estimated to be very low for all scenarios that include large-scale implementation of CO_2 capture in 2030, i.e. a reduction varying between 27% and 41%. This holds especially for the scenario with a large share of oxyfuel combustion technology.

Further, it was found that NO_x emissions from the P&H sector could be 15% higher in a scenario with predominantly post-combustion CO_2 capture compared to the baseline scenario without CCS. A reduction in NO_x emissions is expected when oxyfuel combustion (-16%) or IGCC with pre-combustion CO_2 capture (-20%) is mainly applied.

Large-scale implementation of the post-combustion technology in 2030 may also result in significant higher NH_3 emissions compared to scenarios without CCS and with other CO_2 capture options, although uncertainty in this estimate is substantial. If these emissions are not controlled properly, NH_3emissions from the P&H sector change from an insignificant contribution of 0.5% towards a possible very significant contribution of 13% of the EU total for all sectors together.

Direct particulate matter emissions are likely to be lower in the scenarios with CO_2 capture. The scenario with implementation of the oxyfuel combustion technology shows the largest (i.e. 59%) reduction in PM emissions in the P&H sector followed by the scenario with a significant share allocated to pre-combustion CO_2 capture showing a reduction of 31%. Post–combustion capture may show an increase in PM emissions due to a limited removal and a larger increase in primary energy use. The scenario with post–combustion capture resulted in PM emissions varying between 35% reduction and 26% increase. No robust conclusions could however be drawn on how CO_2 capture influences the emissions of various PM size categories (i.e. $PM_{2.5}$, PM_{10} and >PM_{10}) as this is not satisfactorily addressed in pertaining literature.

6. SPECIAL HIGHLIGHT TOPIC: ATMOSPHERIC EMISSIONS FROM POST-COMBUSTION CAPTURE

6.1. Solvent Emissions

Due to the low partial CO_2 pressure in flue gases from the power sector, the use of chemical solvents is preferred in post-combustion capture. Chemical solvents seem to be the preferred option for the short-term, since this technology is relatively mature, commercially available at industrial scale (though not yet power plant scale), and post-combustion can be used to retrofit existing power stations (as end-of-pipe treatment). The disadvantages of using amines as chemical solvent are high costs for energy (energy penalty), space (due to large gas volumes) and equipment. Furthermore, amines and degradation products are found to be emitted from the stack, causing potential environmental impacts (Horssen, Kuramochi et al. 2009).

Amines can leave the power plant with the CO_2 captured gas to be stored in the deep underground where it is considered to have limited environmental impact. Emissions of amines to the air can take place when the residuals are taken out of process when recycling the amine at the top of the absorber (especially relevant for safety of workers) and with the cleaned flue gas at the top of the absorber where they are emitted into the atmosphere (relevant for the public and environment).

Finally, amines can degrade, e.g. into ammonia (also treated in the previous sections). The amines can also react with oxidized nitrogen in the atmosphere to form potentially harmful compounds such as nitrosamines, nitramines, aldehydes and amides. The environmental impacts are not easy to assess since there are a large number of degradation products which not only depend on the degradation mechanisms occurring in the capture process, but also on the type of amines used (Knudsen 2008).

6.2. Environmental Impacts Of Amines

In 2007, the Norwegian Institute for Air Research (NILU) conducted a screening study to understand more about atmospheric amine chemistry and to evaluate the environmental effects of amine emissions and degradation products such as nitrosamines, nitramines, aldehydes and amides.

The amines studied are monoethanolamine (MEA), piperazine, aminomethylpropanol (AMP) and methyldiethanolamine (MDEA). Among these amines, piperazine has been through a thorough evaluation and classification in the EU system. There are several experimental studies available on MEA, but most of them were conducted during 1960s and -70s. For AMP and MDEA the toxicological data are rare. High quality inhalation studies are lacking. For piperazine and MEA indications exist of reproductive and developmental toxicity. In addition, one study suggests similar effects of AMP. None of the amines have been reported to be carcinogenic, but this should also be evaluated further with additional studies (Låg, Andreassen et al. 2009).

Låg et al. concluded that amines themselves are most likely causing little risk to human health, but the emissions contribute to the nitrogen load and potentially to eutrophication which could have impacts on sensitive terrestrial ecosystems.

Nitrosamines (N-nitrosamines) are a large and diverse family of synthetic and naturally occurring compounds described by the formula (R1)(R2) N-N=O, where R1 and R2 is an alkyl or aryl group. Nitrosamines are typically liquids, oils or volatile solids. Nitrosamines occur in the diet, through use of tobacco, cosmetics, pharmaceutical products and agricultural chemicals. Nearly all commercially available alkylamines are generally contaminated by small quantities of their corresponding N-nitroso analogues. Industrial installations producing or using amines might be a source of nitrosamine pollution (Tricker, Spiegelhalder et al. 1989).

Exact data on concentration levels in power plants using amine based carbon capture are very sparse and very hard to find in the public literature. However, nitrosamines are considered of particular concern because of their toxic and carcinogenic properties at extremely low levels.

Nitramines are also of concern as they are suspected to be carcinogenic, though considerably less than the nitrosamines. However, the longer lifetime in the atmosphere may lead to higher exposure values. Modelling also indicates that amines lower the surface tension of water droplets, which under appropriate climatic conditions can be a trigger for rain with the potential of causing negative impacts to the local environment. Worst case studies for a generic full scale amine plant with conditions representing the west coast of Norway show that the predicted concentrations of photo oxidation compounds are at the same level as the proposed "safety limits", implying that risks to human health and natural environment cannot be ruled out (Knudsen 2009).

With regard to aldehydes, Låg et al. (2009) concluded that at airborne levels for which the prevalence of sensory irritation is minimal, both in incidence and degree (<1.2 mg/m^3), risks of respiratory tract cancer are considered to be negligibly low. Acetamide may induce skin irritation. The irritating potential of the aldehydes and amides might in this context be the most relevant adverse health effect of these compounds, as the amines probably to be used in CO_2 capture also have such effects. Therefore, all these compounds have to be evaluated together with respect to irritating potential of the air around the gas plants (Marit, Instanes et al. 2009).

These statements highlight the necessity for further testing and analysis of amine effects in order to limit the risks, especially for nitrosamines and nitramines. At least as important is the measurement of amine related substances from CCS equipped power plants in order to assess the exact concentration level of specific amine related species. This is needed to assess the risks for public and workers as well as to understand the chemical formation processes as a basis for the development of countermeasures against amine effects due to CCS.

6.3. Co₂ Capture Solvents and Regulations

The health and environmental properties of a number of CO_2 capture compounds have been evaluated by StatoilHydro in the light of the REACH regulation. REACH is the new chemical legislation in the EU. REACH stands for Registration, Evaluation and Authorisation of CHemicals. Industrial CO_2capture plants are covered by REACH and the IPPC, the EU directive restricting polluting discharges from industry. An important item is the discharge permission based on the comparison with the Best Available Technologies (BAT). The current BREF (Reference Document on Best Available Techniques) for large combustion Plants (IPPC 2006) does not contain information on solvent related emissions from CO_2 capture.

Svanes (2008) shows in his study that the selected compounds (mainly amines) and degradation products (ammonia) are not on the restricted list. Most of the compounds are classified as harmful to health and/or the environment. Using the compounds will not be severely restricted by REACH. The study did not incorporate the degradation products as nitrosamines. A more comprehensive study is recommended containing exposure studies and mapping of degradation products.

Based on the available literature Låg et al. (2009) suggested exposure guidelines for four amines; particularly for AMP and MDEA there are few high quality studies. The guidelines presented are therefore just indicative. The uncertainty factors were chosen in accordance with EU guidelines. Based on inhalation exposure risk, the general population, over time, should not be exposed to levels in the air higher than:

- MEA:10μg/m^3
- AMP:6μg/m^3
- MDEA:120μg/m^3
- Piperazine:5μg/m^3

Finally, it has been stated that it is highly relevant to know which precise amine is used in CCS, because each individual amine has different effects and potential risks. Furthermore, use of more than one amine infers that the exposure guidelines should be evaluated again, since amines seem to have similar adverse effects and might therefore also show additive or synergistic effects.

7. CONCLUSIONS AND WAY FORWARD

Depending on the applied CO_2 capture technology, trade-offs and synergies can be expected for key atmospheric emissions, being: NO_x, SO_2, NH_3, particulate matter, Hg, HF and HCl. For all three (pre-, post- and oxyfuel combustion) capture systems it was found that SO_2, NO_x and PM emissions are expected to be reduced or remain equal per unit of primary energy input compared to power plants without CO_2 capture. Increase in primary energy input as a result of the energy penalty for CO_2 capture may for some technologies and substances result in a net increase of emissions per kWh output. The largest increase is found for the emission of NO_x and NH_3 when equipping power plants with post-combustion capture. A decrease is expected for SO_2 emissions, which are low for all power plants with CO_2 capture.

Additional research (measurements and modelling) and regulatory efforts (norm setting) are required to cope with 'new' emissions from predominantly post-combustion CO_2 capture technologies. Laboratory and field experiments are necessary to obtain more precision in the estimates of emission levels, as little information exists in open literature. For this, accurate sampling and analysis methods have to be developed and validated for low concentrations.

Also, for post-combustion capture using amines it is necessary to identify and quantify the specific compounds that will be emitted or formed post-emission, where particular focus should be put on nitrosamines and nitramines. It is recommended to focus research on the determination of atmospheric degradation paths, precise degradation yields, and degradation products' lifetime in the atmosphere. Development of models is necessary to quantify the mass fluxes and chemical interactions, and finally to integrate them in a dispersion model to quantify the load and possible environmental consequences.

Furthermore, research should be focused on the assessment of toxicity levels of these substances, as a basis for the development of both acute and chronic human toxicity exposure limits for amines and associated substances, both for workers and the public.

This is needed to further compile data and information to create a relative ranking of amines with respect to potential environmental and health effects and toxicity and to find efficient ways to mitigate formation of nitrosamines and nitramines.

We recommend to set up extensive environmental monitoring programmes at currently planned CO_2capture (demonstration) plants aimed at creating a better understanding of the formation and fate of solid, liquid and atmospheric pollutants. Emissions that should be monitored are: SO_x, NO_x, HF, HCl, Hg, PAH, dioxins, hydrocarbons, heavy metals, NH_3, MEA and PM. For particulate matter it is especially of interest to discern the removal efficiencies for the various sizes of particulate matter. For heavy metals it is of interest to measure to what extent the transposition occurs from atmospheric emission to waste water effluent and solid waste. Monitoring programmes should help to quantify emissions in further detail and share its knowledge with the wide research community.

Life cycle effects of implementing CO_2 capture options should not be neglected when reviewing the environmental performance of complete CCS chains, from cradle to grave. Recent studies namely indicate that for some substances (e.g. SO_x) direct (atmospheric) emissions may decrease due to CO_2capture; but that additional life cycle emissions by up- and downstream process may result in a deterioration of the overall environmental performance of the CCS chain compared to a power plant without CCS; except for the global warming potential.

For a number of environmental impact categories no agreement exists on the exact direction (positive, negative) or the level of the life cycle impacts due to CCS deployment. The sometimes large ranges often indicate that specific regional or technical issues influence the overall environmental performance of a chain. This requires further research.

We also see a high value in screening next generation CO_2 capture technologies at an early stage on their environmental performance in order to facilitate the optimization of CCS chains not only on its cost and GHG emission performance, but also taking into account other environmental themes.

NOTES

[1] - During field tests MEA (monoethanolamine) and the amine based 'Castor 1' and 'Castor 2' solvents were tested. The Castor 2 solvent resulted in the 40 % uptake of sulphur compared to 80% for MEA.

[2] - This additional reduction succeeding the conventional FGD is achieved by the reaction of SOx with caustic soda in the flue gas cooler which cools the flue gas before it enters the absorber.

[3] - Tests in a research facility indicate that SOx removal was improved in the case of oxygen rich combustion, which can partly be explained by longer gas residence time in the FGD (Marin and Carty 2002; Chatel-Pelage, Marin et al. 2003; Chen, Liu et al. 2007; WRI 2007).

[4] - The main fraction of NOx is formed by NO which is expected to be unaffected by the CO2 capture process. NO2 fraction of NOx, which is typically about 5-10%, may react with the solvent resulting in a reduction of NOx emission per MJprimary. However, also not all of the NO2 is expected to react, i.e. only 25 % (Rao and Rubin 2002; IPCC 2005).

[5] - S(N)CR = Selective (Non) Catalytic Reduction; a technology to reduce NOx emissions by converting NOx into N2 with the use of reactants, such as fun instance ammonia or urea.

[6] - These reactions are (taken from (White, Torrente-Murciano et al. 2008)): 2 NO2 + H2O ↔ HNO2 + HNO3 and 3 HNO2 ↔ HNO3 + 2 NO + H2O.

[7] - A potential by-product of this process may be mercuric nitrate (NO3)-2Hg2+ which is formed due to a reaction between the nitric acid and mercury in the flue gas. Although this substance is highly toxic it means that mercury is effectively removed from the flue gas.

[8] - Particulate matter can be subdivided into particles with a diameter larger than 10 microns (>PM10) and smaller than 10 microns (PM10). PM10 can then be further subdivided into the size categories 'Coarse' (PM2.5-10) and 'fine' (PM2.5).

[9] - 'Dust' (not further specified as PM10 or PM2.5) emissions are reduced by 40-50% in the flue gas cooler prior to the absorption process in which another 40-60% of the particulates is removed from the flue gas.

[10] - They do not report a quantitative estimate but suggest that the flue gas cooler will result in a deep reduction of particulate matter entering the absorption process.

[11] - The efficiency of the Electrostatic Precipitator is possibly improved as a larger share of SOx is represented by SO3 (Tan, Croiset et al. 2006).

[12] - In this case the Piperazine, an amine, is added to the potassium carbonate sorbent as an activator to increase reaction rate.

[13] - Chemical looping concepts typically include an oxidizing reactor (OX) were a metal oxide (oxygen carrying metals that are considered are: Cu, Co, Ni, Fe and Mn) is formed through the exothermic reaction of a metal with oxygen. This is the oxygen carrier that transports the

oxygen to the reduction reactor (RED). In the reduction reactor the fuel reacts (oxidizes) with the oxygen from the metal oxide (MeO).

[14] - NOx and SOx emissions lead to the formation of acid gases, which can lead to acidification, eutrophication and smog formation nd not in the consequence of a given emission. For instance, SO2 and NOx can cause acidification.

[15] - Two outliers are identified. Markevitz et al., (2009) shows a significantly smaller (23%) share of power plant operation after capture while Pehnt and Henkel (2009) showed a larger share (79%). The origin of the differences cannot be identified from the data reported.

[16] - From 2010 onwards it is more likely that the power plants will be built capture ready. In this study it was therefore assumed that only power plants built between 2010 and 2020 are retrofitted between 2020 and 2030.

[17] - Greenhouse gas - Air pollution Interactions and Synergies model. The GAINS model is developed to analyse trade-offs and co-benefits of strategies aimed at the reduction of air pollution and greenhouse gases on the medium-term, i.e. until 2030.

REFERENCES

1. M. Akai, et al. 1997 "Life-cycle analysis of a fossil-fuel power plant with CO2 recovery and a sequestering system." Energy 22(2-3): 249 EOF255 EOF .

2. S. Allaie, D. Jaspers, 2008 TNO CATO pilot plant at E-On Maasvlakte (confidential report). Delft, the Netherlands, TNO: 29

3. K. Andersson, 2007 Combustion tests and Modelling of the oxy-fuel process-an overview of research activities at Chalmers University, Sweden. 2nd International Oxy-Combustion Network Meeting Windsor, Connecticut, USA IEA GHG, Alstom Power.

4. K. Andersson, F. Johnsson, 2006 "Process evaluation of an 865 MWe lignite fired O2/CO2 power plant." Energy Conversion and Management 47(18-19): 3487.

5. M. Anheden, et al. 2005 "Denitrogenation (or Oxyfuel Concepts)." Oil & Gas Science and Technology Rev. IFP 60 3 485495 .

6. Aspelund, et al. 2006 "Ship transport of CO2 - Technical solutions and analysis of costs, energy utilization, exergy efficiency and CO2 emissions." Chemical Engineering Research & Design 84(A9): 847-855

7. Bauer, T. Heck, 2009 Carbon capture and storage: Life cycle assessment and external costs of future fossil power generation.. International Conference on Life Cycle Management 2009.

8. J. Bradshaw, et al. 2006 CO2 storage capacity estimation: issues and development of standards. Greenhouse Gas Control Technologies 8, Trondheim.

9. B. J. P. Buhre, et al. 2005 "Oxy-fuel combustion technology for coal-fired power generation." Progress in Energy and Combustion Science 31(4): 283 EOF307 EOF .

10. U. Burchhardt, 2009 Experiences from Commissioning and Test Operation of Vattenfall's Oxyfuel Pilot Plant. 1st International Oxyfuel Combustion Conference, Cottbus, Germany.

11. F. Chatel-Pelage, et al. 2003 A pilot-scale demonstration of oxy-combustion with flue gas recirculation in a pulverized coal-fired boiler. 28th international conference on coal utilization & fuel systems- March 2003, Clearwater, Florida,.

12. J. Chen, C. , et al. 2007 "Emission characteristics of coal combustion in different O2/N2, O2/CO2 and O2/RFG atmosphere." Journal of Hazardous Materials 142(1-2): 266.

13. P. Chiesa, et al. 2005 "Using Hydrogen as Gas Turbine Fuel." Journal of Engineering for Gas Turbines and Power 127 7380 .

14. Corti, L. Lombardi, 2004 "Reduction of carbon dioxide emissions from a SCGT/CC by ammonia solution absorption- preliminary results." International Journal of Thermodynamics 7 4 173181 .

15. E. Croiset, K. V. Thambimuthu, 2001 "NOx and SO2 emissions from O2/CO2 recycle coal combustion." Fuel 80(14): 2117 EOF .

16. Z. Cui, et al. 2010 "Simultaneous Capture of Mercury and CO2 in Amine-Based CO2 Absorption Process." Industrial & Engineering Chemistry Research 49 24 1257612586 .

17. K. Damen, et al. 2006 "A comparison of electricity and hydrogen production systems with CO2 capture and storage. Part A: Review and selection of promising conversion and capture technologies." Progress in Energy and Combustion Science 32(2): 215 EOF246 EOF.

18. J. Davison, 2007 "Performance and costs of power plants with capture and storage of CO2." Energy 32(7): 1163 EOF1176 EOF .

19. DOE and NETL 2007 Pulverized Coal Oxycombustion Power Plants- 1 Bituminous Coal to Electricity.

20. DOE/NETL 2007 Cost and Performance Baseline for Fossil Energy Plants- 1 Bituminous Coal and Natural Gas to Electricity, National Energy Technology Laboratory, US department of Energy.

21. S. Ertesvag, et al. 5 EOF39 EOF 2005 "Exergy analysis of a gas-turbine combined-cycle power plant with precombustion CO2 capture." Energy 30(1): 5.

22. H. Farzan, et al. 2005 Pilot-Scale Evaluation of Coal Combustion in an Oxygen-Enriched Recycled Flue Gas. The 30th International Conference on Coal Utilization and Fuel Systems, Clearwater, Florida, U.S.A.

23. T. v. Harmelen, et al. 2008 The impacts of CO2 capture technologies on transboundary air pollution in the Netherlands. Utrecht, the Netherlands, TNO, Group Science, Technology and Society

24. E. Hertwich, et al. 2008 "Life-cycle assessment of carbon dioxide capture for enhanced oil recovery." Chinese Journal of Chemical Engineering 16 3 343353

25. Hetland, T. Christensen, 2008 "Assessment of a fully integrated SARGAS process operating on coal with near zero emissions." Applied Thermal Engineering 28 16 20302038 .

26. Hopman. 2008 Personal Communication. TNO, Delft

27. v. Horssen, et al. 2009 The impacts of CO2 capture technologies in power generation and industry on greenhouse gases emissions and air pollutants in the Netherlands. Dutch Policy Research Programme on Air and Climate (BOLK). Utrecht, Utrecht University, TNO: 160

28. IEA GHG 2004 Ship transport of CO2. Cheltenham, International Energy Agency Greenhouse Gas R&D Programme

29. IEA GHG 2006a Environmental impact of solvent scrubbing of CO2 R. v. Gijlswijk, et al, TNO, IEA Greenhouse Gas R&D Programme

30. IEA GHG 2006b international Oxy combustion Research Network for CO2 capture; Report on Inaugural workshop

31. IEA GHG 2006c Near zero emission technology for CO2 capture from power plant IEA Greenhouse Gas R&D programme

32. IIASA. 2008 "Greenhouse Gas and Air Pollution Interactions and Synergies (GAINS) ".

33. M. Iijima, et al. 2007 "Long-Term Demonstration of CO2 Recovery from the Flue Gas of a Coal-Fired Power Station " Technical review-Mitsubishi Heavy Industries 44(2).

34. IPCC 2005 IPCC Special Report on Carbon Dioxide Capture and Storage. Prepared by Working Group III of the Intergovernmental Panel on Climate Change. B. Metzet al. Cambridge University Press, Cambridge, United Kingdom and New York, NY, USA,: 442.

35. IPPC 2006 Reference Document on Best Available Techniques for Large Combustion Plants, Integrated Pollution Prevention and Control Bureau,.

36. H. H. Khoo, R. B. H. Tan, 2006 "Life Cycle Investigation of CO2 Recovery and Sequestration." Environ. Sci. Technol. 40 12 40164024 .

37. S. Kishimoto, et al. 2008 Current Status of MHI's CO2 Recovery Technology and Optimization of CO2 Recovery Plant with a PC Fired Power Plant. 9th International Conference on Greenhouse Gas Control Technologies (GHGT-9), Washington DC, USA.

38. N. Knudsen, et al. 2008 Experience with CO2 capture from coal flue gas in pilot-scale: Testing of different amine solvents. 9th International Conference on Greenhouse Gas Control Technologies (GHGT-9), Washington DC, USA.

39. J. N. Knudsen, et al. 2006 CASTOR 1 t/h CO2 absorption pilot plant at the Elsam Kraft A/S Esbjerg power plant- First year operation experience. Greenhouse Gas Control Technologies 8, Trondheim, Norway.

40. S. Knudsen, et al. 2009 Summary Report: Amine Emissions to Air during Carbon Capture, Phase I: CO2 and Amines Screening Study for Effects to the Environment, NILU report N-108068

41. S. Knudsen, et al. 2008 Environmental impact of amines from CO2 capture, NILU report O-107124.

42. H. Knuutila, et al. 2009 "CO2 capture from coal-fired power plants based on sodium carbonate slurry; a systems feasibility and

sensitivity study." International Journal of Greenhouse Gas Control 3(2): 143 EOF151 EOF .

43. J. Koornneef, et al. 2010 "The impact of CO2 capture in the power and heat sector on the emission of SO2, NOx, Particulate Matter, Volatile Organic Compounds and NH3 in the European Union." Atmospheric Environment 44 13691385

44. J. Koornneef, et al. 2008 "Life cycle assessment of a pulverized coal power plant with post-combustion capture, transport and storage of CO2." International Journal of Greenhouse Gas Control 2(4): 448 EOF467 EOF .

45. Korre, et al. 2010 "Life cycle modelling of fossil fuel power generation with post-combustion CO2 capture." International Journal of Greenhouse Gas Control 4 289300 .

46. Korre, et al.press. (in, . , "Life cycle modelling of fossil fuel power generation with post-combustion CO2 capture." International Journal of Greenhouse Gas Control In Press, Corrected Proof.

47. F. Kozak, et al. 2008 Chilled Ammonia Process for CO2 Capture. 9th International Conference on Greenhouse Gas Control Technologies (GHGT-9), Washington DC, USA.

48. H. M. Kvamsdal, et al. 2007 "A quantitative comparison of gas turbine cycles with CO2 capture." Energy(32): 10 EOF24 EOF .

49. H. M. Kvamsdal, T. Mejdell, 2005 Tjeldberodden power/methanol-CO2 reduction efforts SP2: CO2 capture and transport, Sintef energy research

50. Låg, et al. 2009 Health effects of different amines relevant for CO2 capture, NILU report N-108068

51. Lombardi, 2003 "Life cycle assessment comparison of technical solutions for CO2 emissions reduction in power generation." Energy Conversion and Management 44(1): 93 EOF108 EOF .

52. Anheden, et al. 2005 "Denitrogenation (or Oxyfuel Concepts)." Oil & Gas Science and Technology Rev. IFP 60 3 485495 .

53. Marin, R. Carty, 2002 Demonstration study of high sulfur coal combustion in oxygen enriched flue gas Air Liquide, Illinois Clean Coal Institute.

54. L. Marit, et al. 2009 Health effects of possible degradation products of different amines relevant for the CO2 capture, NILU report N-108068

55. Markewitz, et al. 2009 "Environmental impacts of a Germand CCS strategy." Energy Procedia 1 37633770 .

56. Moser, S. Schmidt, K. Stahl, 2011 Investigation of trace elements in the inlet and outlet streams of a MEA-based post-combustion capture process Results from the test programme at the Niederaussem pilot plant. Greenhouse Gas Control Technologies 10, Amsterdam.

57. NEEDS 2009 NEEDS: New Energy Externalities Developments for Sustainability. Deliverable 6 1 External costs from emerging electricity generation technologies, Sixth Framework Programme 2009

58. Nexant Inc. 2006 Environmental Footprints and Costs of Coal-Based Integrated Gasification Combined Cycle and Pulverized Coal Technologies. Washington, U.S. Environmental Protection Agency.

59. Z. Nie, 2009 Life cycle modeling of carbon dioxide capture and geological storage in energy production. PhD thesis Imperial College London

60. NUON 2005 Milieujaarverslag 2004 Nuon Power Buggenum B.V.- Willem-Alexander Centrale

61. NUON 2006 Milieujaarverslag 2005 Nuon Power Buggenum B.V.- Willem-Alexander Centrale

62. N. A. Odeh, T. T. Cockerill, 2008 "Life cycle GHG assessment of fossil fuel power plants with carbon capture and storage." Energy Policy 36(1): 367 EOF380 EOF .

63. J. Oexmann, A. Kather, 2009 "Post-combustion CO2 capture in coal-fired power plants: Comparison of integrated chemical absorption processes with piperazine promoted potassium carbonate and MEA." Energy Procedia 1 1 799806 .

64. M. Pehnt, J. Henkel, 2009 "Life cycle assessment of carbon dioxide capture and storage from lignite power plants." International Journal of Greenhouse Gas Control 3(1): 49 EOF66 EOF .

65. Rao, 2006 Implications of CO2 Sequestration for Gas Turbines-section 1.2.2 The Gas Turbine Handbook R. Dennis, U.S. Department of Energy, National Energy Technology Laboratory.

66. B. Rao, E. S. Rubin, 2002 "A technical, economic, and environmental assessment of amine-based CO2 capture technology for power plant greenhouse gas control." Environmental Science and Technology 36(20): 4467.

67. M. Rao, et al. 2004 An integrated modeling framework for carbon management technologies. 1-Technical documentation: Amine-based CO2 capture and storage systems for fossil fuel power plant. Pittsburg, United States, Carnegie Mellon University.

68. RECCS 2008 RECCS: Ecological, economic and structural comparison of renewable energy technologies with carbon capture and storage- an integrated approach, Federal Ministry for the Environment, Nature Conservation and Nuclear Safety (BMU)

69. Schreiber, et al. 2009 "Environmental assessment of German Electricity generation from coal-fired power plants with amine-based carbon capture." International Journal of Life Cycle Assessment 14 547559

70. K. Smith, et al. 2009 "Recent developments in solvent absorption technologies at the CO2CRC in Australia." Energy Procedia 1 1 15491555 .

71. P. Spath, M. Mann, 2004 Biomass power and conventional fossil systems with and without CO2 sequestration-comparing the energy balance, greenhouse gas emissions and economics. Golden, Colorado, National Renewable Energy Laboratory: 28 p.

72. R. Strazisar, et al. 2003 "Degradation Pathways for Monoethanolamine in a CO2 Capture Facility." Energy Fuels 17 4 10341039 .

73. G. Sundkvist, et al. 2004 AZEP gas turbine combined cylce power plants-thermal optimisation and LCA analysis. 7th International Conference on Greenhouse Gas Control Technologies, Vancouver, Canada.

74. Supap, et al. 2009 "Kinetics of sulfur dioxide- and oxygen-induced degradation of aqueous monoethanolamine solution during CO2 absorption from power plant flue gas streams." International Journal of Greenhouse Gas Control 3 2 133142 .

75. E. Svanes, 2008 Evaluation of direct toxic impact of CO2 capture chemicals in view of REACH regulation, Østfoldforskning report

76. Y. Tan, et al. 2006 "Combustion characteristics of coal in a mixture of oxygen and recycled flue gas." Fuel 85(4): 507 EOF512 EOF .

77. Thébault, et al. 2009 Behaviors of NOX and SOX in CO2 Compression and Purification Processes- Experience at 30 MWth Oxy-Coal Combustion CO2 capture Pilot Plant. 1st Oxyfuel Combustion Conference. Cottbus, Germany

78. Thitakamol, et al. 2007 "Environmental impacts of absorption-based CO2 capture unit for post-combustion treatment of flue gas from coal-fired power plant." International Journal of Greenhouse Gas Control 1(3): 318 EOF342 EOF .

79. C. Thomas, S. M. Benson, 2005 Carbon dioxide capture for storage in deep geological formations- Results from the CO2 capture project. 2 Geologic storage of carbon dioxide with monitoring and verification. Amsterdam, Elsevier.

80. R. Tricker, et al. 1989 "Environmental exposure to preformed nitroso compounds " Canc. Surv. 8 251272

81. Tzimas, et al. 2007 "Trade-off in emissions of acid gas pollutants and of carbon dioxide in fossil fuel power plants with carbon capture." Energy Policy 35(8): 3991 EOF3998 EOF .

82. Van Bergen, et al. 2003 Feasibility study on CO2 sequestration and enhanced CBM production in Zuid-Limburg, Novem: 76.

83. P. Viebahn, et al. 2007 "Comparison of carbon capture and storage with renewable energy technologies regarding structural, economic, and ecological aspects in Germany." International Journal of Greenhouse Gas Control 1(1): 121 EOF133 EOF .

84. H. Waku, et al. 1995 "Life cycle analysis of fossil power plant with CO2 recovery and sequestering system." Energy Conversion and Management 36(6-9): 877 EOF880 EOF .

85. Weisser, 2007 " A guide to life-cycle greenhouse gas (GHG) emissions from electric supply technologies." Energy 32 15431559

86. V. White, et al. 2006 Purification of oxyfuel-derived CO2 for sequestration for EOR. Greenhouse Gas Control Technologies 8, Trondheim.

87. V. White, et al. 2008 Purification of Oxyfuel-Derived CO2. 9th International Conference on Greenhouse Gas Control Technologies (GHGT-9), Washington DC, USA.

88. WRI 2007 Development of a Novel Oxygen Supply Process and its Integration with an Oxy-Fuel Coal-Fired Boiler. Laramie, Wyoming, Western Research Institute.

89. Yamada, 2007 Pilot scale experiments giving direct comparison between air and oxy-firing of coals and implication for large scale plant design; IHI, Japan.. 2nd International Oxy-Combustion Network Meeting Windsor, Connecticut, USA IEA GHG, Alstom Power.

90. J. Yan, et al. 2006 Conceptual Development of Flue Gas Cleaning for CO2 Capture from Coal-fired Oxyfuel Combustion Power Plant. Greenhouse Gas Control Technologies 8, Trondheim.

91. J. Yan, et al. 2009 Flue-gas Cleaning Processes for CO2 Capture from Oxyfuel Combustion- Experience of FGD and FGC at 30 MWth Oxyfuel Combustion Plant. 1st Oxyfuel Combustion Conference. Cottbus, Germany

92. C. Yeh, H. Bai, 1999 "Comparison of ammonia and monoethanolamine solvents to reduce CO2 greenhouse gas emissions." The Science of The Total Environment 228(2-3): 121 EOF133 EOF

93. ZEP 2006 Power Plant and Carbon Dioxide Capture-The final report from Working Group 1, The European Technology Platform for Zero Emission Fossil Fuel Power Plants (ZEP)

94. Z. X. Zhang, et al. 2006 "Optimization of pipeline transport for CO2 sequestration." Energy Conversion and Management 47(6): 702 EOF715 EOF .

CHAPTER 2

Al2O3-enhanced Macro/Mesoporous Fe/TiO2 for Breaking Down Nitric Oxide

Dieqing Zhang[1] and Guisheng Li[1]

[1] Department of Chemistry, Key Laboratory of Resource Chemistry of Ministry of Education, Shanghai Normal University, Shanghai 200234, China

1. INTRODUCTION

High air contaminant levels in the indoor environment come from either the ambient air or from indoor sources. (Cao, 2001) Nitrogen oxide is one of the most common gaseous pollutants found in the indoor environment with the concentration in the range of 70-500 parts-per-billion (ppb) levels. This has serious implications on the environment and health of the mankind. (Huang et al., 2009) Conventional techniques to treat nitric oxide in industrial emission mainly include physical adsorption, biofiltration, and thermal catalysis methods. However, these methods usually suffer from some disadvantages, such as the low efficiency for pollutants at the parts per billion level and the difficulty in solving the postdisposal and regeneration problems. (Huang et al., 2008)

As a promising environmental remediation technology, semiconductor-mediated photocatalytic technology has been widely used to purify contaminated air and wastewater. (Fox & Dulay, 1993) Titanium dioxide is the most widely used photocatalyst because of its superior photoreactivity, nontoxicity, long-term stability and low price. Recently, great attention has been paid to macro/mesoporous TiO_2 for its interconnected macroporous and mesoporous structures. Such hierarchical material may enhance properties compared with single-sized pore materials due to increased mass transport through the material and minimized pressure drop over the monolithic material.(Yuan et al. 2006) Meanwhile the macroporous channels could serve as light-transfer paths for the distribution of photon energy onto the large surface of inner photoactive mesoporous frameworks. Therefore, higher light utilization efficiency could be obtained for heterogeneous photocatalytic systems including photooxidation degradation and solar cells. In addition, the hierarchical structure-in-structure arrangement of mesopore and macropore is benefit for the molecule traffic control and for the resistance of the photocatalyst to poisoning by inert deposits.(Rolison 2003)

Though such structure contributes great advantages to TiO_2, such as a readily accessible pore-wall system and better transport of matter compared to the traditional TiO_2 photocatalysts, the anatase TiO_2 semiconductor has a relatively large band gap of 3.2 eV, corresponding to a wavelength of 388 nm.(Yu et al., 2006) The requirement of UV excitation impedes the development of solar-driven photocatalytic systems. As a promising way, doping method can effectively extend the light absorption of TiO_2 to the visible region and reduce the recombination of photoinduced electrons and holes.(Zhu et al., 2007) Among various dopants, the Fe^{3+}-dopant is most frequently employed owing to its unique half-filled electronic configuration, which might narrow the energy gap through the formation of new intermediate energy levels and also diminish recombination of photoinduced electrons and holes by capturing photoelectrons. However, calcination of the photocatalysts at high temperature is usually indispensable for removing organic templates, enhancing structural crystallization, and allowing doped ions to enter into the frameworks of TiO_2.(Wang et al., 2009) Such treatment at high temperature will result in great loss of surface area and destroying the pore systems owing to the grain growth, especially for porous materials. Thus, the photoactivity of the calcined samples with low specific area will be greatly reduced for the poor light-harvesting capability.(Yu et al., 2006) Fortunately, using inorganic structure stabilizers (SiO_2, ZrO_2, and Al_2O_3) could allow the anti-sintering properties of porous materials to be promoted greatly enough for application in high temperature environment, such as treating automotive exhaust.(Wang et al., 1999)

In this chapter, we describe a detailed study of the effect of Al_2O_3 as a promoter in enhancing a macro/mesoporous visible-light photocatalyst, Fe/TiO_2, for the oxidation of nitric oxide (NO). The photocatalysts are synthesized through directing the formation of inorganic phases (Al_2O_3-Fe/TiO_2) with multidimensional pore systems through the self-assembly of a single surfactant under hydrothermal conditions. The experimental results showed that doping Fe^{3+} into the framework of TiO_2 can effectively extend the optical absorption spectrum to visible light range. Introducing highly dispersed amorphous Al_2O_3 species into the Fe/TiO_2 system could greatly increased the thermal stability of the Fe/TiO_2 framework with higher surface area and larger pore volume. It is surprising that the Al_2O_3-Fe/TiO_2 sample treated at 700° C possessed a high specific surface area (ca. 130 m^2/g), about 6 times of that of the Al_2O_3-free sample. The photooxidation of NO in air over the 3D macro/mesoporous Al_2O_3-Fe/TiO_2 photocatalysts was studied. These products were utilized to remove gaseous NO at 400 parts-per-billion level in air under visible-light irradiation. These Al_2O_3-Fe/TiO_2photocatalysts exhibited very strong ability to oxidize the NO gas in air under visible-light irradiation. Importantly, these 3D macro/mesoporous Al_2O_3-Fe/TiO_2 photocatalysts showed excellent stability and maintained a high level of photocatalytic activity after multiple reaction cycles.

2. EXPERIMENT SECTION

2.1. Preparation Of 3d Macro/Mesoporous Al_2o_3-Fe/Tio_2 Photocataltsts

Brij 56 [$C_{16}(EO)_{10}$], titanium isopropoxide, aluminum sec-butoxide, and ferric (III) nitrate are purchased from Aldrich. All chemicals were used as received. In a typical synthesis of macro/mesoporous visible light photocatalysts Al_2O_3-Fe/TiO_2, required amount of ferric (III) nitrate was dissolved in a aqueous solution of Brij 56 (15 wt %) with pH = 2 adjusted by sulfuric acid under ultrasonic irradiation in an ultrasonic clean bath (Bransonic ultrasonic cleaner, model 3210E DTH, 47 kHz, 120 W, USA). 18 ml mixture of aluminum sec-butoxide and titanium isopropoxide with a metal-to-metal molar ratio (M_{Al}/M_{Ti} = 20:100) was added drop by drop into the above medium under stirring, followed by further stirring for 0.5 h. The obtained mixture was then transferred to a Teflon-lined autoclave and heated at 80° C for 36 h under static condition during which the inorganic precursor hydrolyses and polymerizes into a metal oxide network. Finally, the as-prepared white samples were clacined at 400 700° C for 8 h at 1 C/min to remove the surfactant species and improve the crystallinity. The as-prepared Al_2O_3-Fe/TiO_2 samples were denoted as Al-Fe/TiO_2-400, Al-Fe/TiO_2-500, Al-Fe/TiO_2-600 and Al-Fe/TiO_2-700, where 400-700 refers to the calcinations temperature. For comparation, macro/mesoporous photocataltsts, pure TiO_2 and Fe/TiO_2, were also prepared by the same procedure. The molar ratio of Fe/Ti is 0.25 % for all the Fe doped samples.

2.2. Characterization

X-ray diffraction (XRD) measurements were carried out using a Bruker D8 Advance X-ray diffractometer (Cu $K\alpha_1$ irradiation, λ = 1.5406 Å) at a scanning rate of 0.02 Degree/Second. The Schrerrer equation ($\Phi = K\lambda/\beta\cos\theta$) was used to calculate the crystal size.(Machida, Norimoto et al. 1999) In the above equation, λ (0.154 nm) is the wavelength of the X-ray irradiation, K is a constant of 0.89, β is the peak width at half-maximum height after subtraction of the instrumental line broadening using silicon as a standard, and $2\theta = 25.3$ and $2\theta = 27.4$ for anatase and rutile. The phase composition was estimated using the following equations: rutile % = $100 \times (0.884A/R + 1)^{-1}$,(Machida, Norimoto et al. 1999) where A is the peak area of anatase (101) and R is the peak area of rutile (110). The intensity of both of the two peaks is the most intense reflection in the diffractograms. The number of 0.884 is the coefficient of scattering. The morphology and the surface roughness of as-prepared samples were examined by a LEO 1450 VP scanning microscope. Standard transmission electron microscopy images were recorded using a CM-120 microscope (Philips, 120 kV). High-resolution transmission electron microscopy (HRTEM) was recorded in JEOL-2010F at 200 kV. A trace amount of sample was suspended in ethanol solution. After sonication for 10 min, carbon-coated copper grids were used to hold the samples followed by drying. Nitrogen adsorption-desorption isotherms were analyzed at 77 K using Micromeritics ASAP 2010 equipment. The reflectance spectra of the samples over a range of 200-700 nm were recorded by a Varian Cary 100 Scan UV-vis system equipped with a Labsphere diffuse reflectance accessory. Labsphere USRS-99-010 was employed as a reflectance standard. FT-IR spectra on pellets of the samples mixed with KBr were recorded on a Nicolet Magna 560 FT-IR spectrometer.

2.3. Photocatalytic Activity Testing

The photocatalytic experiments for the removal of NO gas in air were performed at ambient temperature in a continuous flow rectangular reactor (10 H cm*30 L cm*15Wcm). A 300Wcommercial tungsten halogen lamp (General Electric) was used as the simulated solar light source. A piece of Pyrex glass was used to cut off the UV light below 400 nm. Four minifans were used to cool theflowsystem. Photocatalyst (0.2 g) was coated onto a dish with a diameter of 12.0 cm. The coated dish was then pretreated at 70° C to remove water in the suspension. The NO gas was acquired from compressed gas cylinder at a concentration of 48 ppm NO (N_2 balance, BOC gas) with traceable National Institute of Stands and Technology (NIST) standard. The initial concentration of NO was diluted to about 400 ppb by the air stream supplied by a zero air generator (Thermo Environmental Inc. model 111). The desired humidity level of the NO flow was controlled at 70% (2100 ppmv) by passing the zero air streams through a humidification chamber. The gas streams were premixed completely by a gas blender and the flow rate was controlled at 4 L.min-1 by a mass flow controller. After the adsorption-desorption equilibrium among water vapor, gases and photocatalysts was achieved, the lamp was turned on. The concentration of NO was continuously measured by a chemiluminescence NO analyzer (Thermo Environmental Instruments Inc. model 42c), with a sampling rate of 0.7 L/min.

3. RESULTS AND DISCUSSION

3.1. X-Ray Diffraction and N_2 Sorption

The crystal composition, thermal stability and mesoporous structure of the as-prepared samples were investigated by X-ray diffraction (XRD) and N_2 sorption analyses. Figure 1a shows the wide-angle XRD patterns of the Fe-doped TiO_2 calcined at different temperatures. For the 400° C sintering sample, a broad peak corresponding to (101) diffraction of anatase-TiO_2 (JCPDS 21-1272) was observed. The broadening of the diffraction peak may have been caused by the small crystalline grain size (6.1 nm). Upon increasing the temperature to 500 C, the intensity of this peak became stronger and sharper, indicating that larger particles (8.0 nm) were formed. However, when the calcination temperature was increased to 600 C, the intensity of the anatase-TiO_2 diffraction peak decreased. Meanwhile, weak peaks indexable as diffractions of rutile-TiO_2 (JCPDS 87-920) appeared. About 4.5 % of the anatase-TiO_2 was converted to rutile-TiO_2.

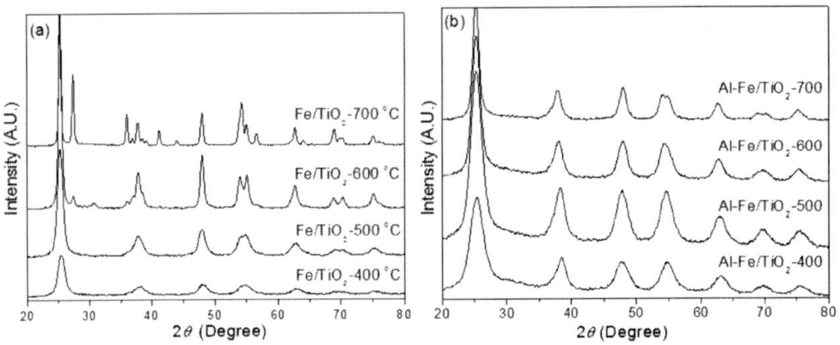

Figure 1. XRD patterns of the as-prepared Fe/TiO_2 and Al-Fe/TiO_2 samples calcined at different temperatures.

After calcinations at 700° C, the size of anatase-TiO_2 increased dramatically to 20.0 nm, and nearly 43.3 % of the anatase-TiO_2 was converted to rutile-TiO_2. Nevertheless, introducing Al_2O_3 species into the Fe-doped TiO_2 system greatly inhibited the crystal growth and the phase-transition. As shown inFigure 1b, all of the Al_2O_3 modified Fe/TiO_2 samples exhibited pure anatase phase at different temperatures. More interestingly, the crystal size of the anatase TiO_2 was greatly decreased after modifying Al_2O_3. Even after 700° C calcination, the grain size can be maintained about 7.2 nm, much smaller than that (20.0 nm) of Fe/TiO_2 sample calcined at 700° C. These wide-angle XRD results revealed that thermal treatment induced the growth of crystal size and subsequent phase transition could be effectively prohibited by doped Al_2O_3 species acting as structural agents. As known, thermal-induced changes in crystal composition and size also had remarkable effects on the textural properties of TiO_2 framework. N_2 sorption analyses were utilized to confirm the change of textural properties.

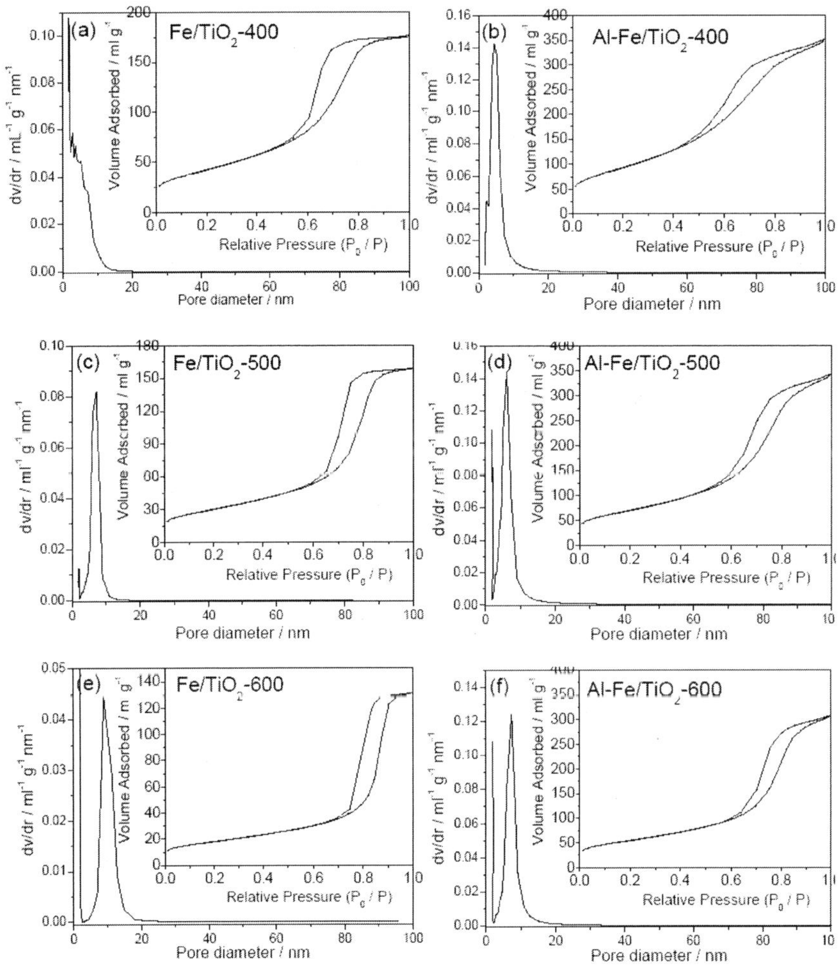

Figure 2. N_2 adsorption-desorption isotherms (a, c, and e) and pore size distributions (b, d, and f) of the as-prepared Fe/TiO_2 and Al-Fe-TiO_2 samples calcined at different temperatures.

N_2 sorption analyses were utilized to investigate the change of textural properties of the as-prepared products. Figure 2 shows the N_2 sorption isotherms and pore size distributions for the modified titanium dioxide calcined at different temperatures. Upon 400° C calcinations, both of Fe/TiO_2-400 and Al-Fe/TiO_2-400 samples exhibited stepwise adsorption and desorption (type IV isotherms) as shown in Figure 2a and b, indicative of a typical mesoporous structure within the as-prepared samples. As shown in Table 1, Al-Fe/TiO_2-400 sample possesses a surface area of 340.0 m^2/g, much higher than that (153.6 m^2/g) of Fe/TiO_2-400 owing to the anti-agglutination effect of induced Al_2O_3 species. The modification of Al_2O_3 also contributed a super lager pore volume (0.53 cm^3/g) to Al-Fe/TiO_2-400 sample. It is about two times of that (0.27 cm^3/g) of Fe/TiO_2-400. Such high surface area and large pore volume will make this material an excellent photocatalyst for its strong adsorption capability. With increase the calcinations temperature, the mesoporous structure of Fe/TiO_2 sample was destroyed. When the calcinations temperature was increased to 700° C, the surface area was decreased to 22.6 m^2/g, and the pore volume was decreased to 0.11 cm^3/g. This is an indication of the collapse of the pore. However, Al-Fe/TiO_2 still owns a high surface area of 131.8 m^2/g and pore volume of 0.44 cm^3/g. To the case of the pore-size distribution, the modification of Al_2O_3 also inhibited the pore size changing owing to its porous structure-stabilizing capability.

BET surface area is calculated from the linear part of the BET plot ($p/p0$ = 0.1-0.2).[b] The total pore volumes are estimated from the adsorbed amount at a relative pressure of $p/p0$ = 0.99.[c] The pore-size distributions (PSD) are derived from the adsorption branches of the isotherms by using the Barrett-Joyner-Halenda (BJH) method.[d] Crystal size was calculated based on XRD results.

3.2. Scanning Electron Microscopy (Sem)

The N_2 sorption analyses could provide mesoporous structure information of the as-prepared materials. To the case of the macroscopic properties, scanning electron microscopy (SEM) should be utilized to examine the macrostructure of the modified TiO_2 monolithic particles. Meanwhile, the high-resolution state of SEM images could also give information on the mesoscopic properties. As shown in Figure 3a,Al-Fe/TiO_2-400 is typically in a large monolithic form (> 30 μm), and exhibits macroscopic network structure with relatively homogeneous macropores of 1~2 μm (size) and about 20 μm (length) in dimension as shown Figure 3b. It is more interesting that these ultralong macroscopic channels are arranged parallel to each other. Figure 3b also demonstrates the extension of the parallel-arrayed macropores completely through the material from the side view of the sample. Such open-ended tubelike macrochannels could serve as ideal light-transport routes for introducing more photoenergy into the interior of the framework of TiO_2. Meanwhile, the high-resolution SEM images (Figure 3c)shows that the walls of the macroporous TiO_2 frameworks are composed of small interconnected TiO_2particles. The mesoporous structure of the as-prepared materials is probably partly due to the intraparticle porosity and partly due to the interparticle porosity of these fine particulates. (Wang, Yu et al. 2005) The macro/mesoporous structure nearly can be maintained even after 600 °C calcinations.

Table 1. Textural properties and crystalline structures and of the prepared porous samples.

Sample	Surface Area/m²/g[a]	Pore Volume/mL/g[b]	Pore Size/nm[c]	Rutile content/Wt %	Crystal Size/nm[d]
Fe/TiO_2-400	153.6	0.27	5.9	0	6.1
Fe/TiO_2-500	109.5	0.24	7.6	0	8
Fe/TiO_2-600	64.9	0.20	11.3	4.5	13.4
Fe/TiO_2-700	22.6	0.11	31.2	43.3	20.0
Al-Fe/TiO_2-400	340.0	0.53	5.3	0	3.2
Al-Fe/TiO_2-500	254.7	0.52	6.8	0	4.0
Al-Fe/TiO_2-600	197.1	0.46	8.1	0	5.3
Al-Fe/TiO_2-700	131.8	0.44	12.4	0	7.2

[a] BET surface area is calculated from the linear part of the BET plot (p/p_0 = 0.1-0.2).[b] The total pore volumes are estimated from the adsorbed amount at a relative pressure of p/p_0 = 0.99.[c] The pore-size distributions (PSD) are derived from the adsorption branches of the isotherms by using the Barrett-Joyner-Halenda (BJH) method.[d] Crystal size was calculated based on XRD results.

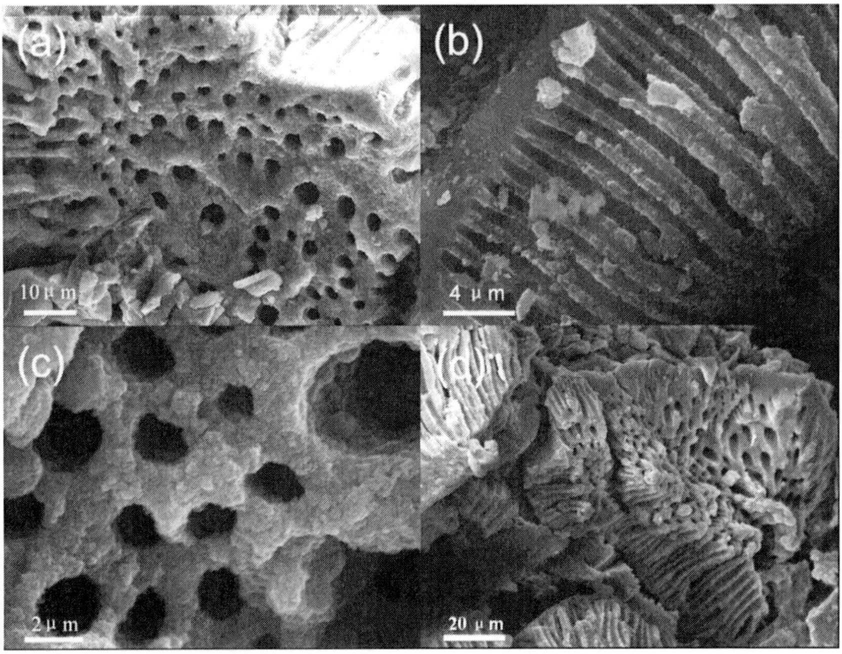

Figure 3. SEM images of the Al-Fe/TiO_2 samples calcined at 400 °C (a, b, c), and 600° C (d).

3.3. Uv-Vis Spectra

UV-visible diffuse reflectance spectroscopy (DRS) was utulized to investigate the electronic states of the as prepared samples. Figure 4a shows the UV-visible absorption spectra of TiO_2-400, Fe/TiO_2-400, Al-Fe/TiO_2-400 and Al-Fe/TiO_2-700 samples. For large energy gap of anatase (3.2 eV), TiO_2-400 sample has no significant absorbance for visible-light. Upon doping Fe^{3+} ions, the light absorption edge of Fe/TiO_2-400 sample was extended to visible light region (< 650 nm) attributed to the formation of Fe-intermediate energy levels, resulting in a decrease in the energy band. The other three samples exhibit a broad absorption bands from 200 to 600 nm with respect to the pure TiO_2, indicating the effective photo-absorption property for this macro/mesoporous structure oxide composite photocatalyst system.

This is because Fe-doping induces the absorbance for visible light owing to, leading to a decrease in the energy band gap.(Nahar, Hasegawa et al. 2006)

Compared to Fe/TiO_2-400 sample, Al-Fe/TiO_2 shows a higher light-absorbance ability located in 200~400 nm. This is because the macro/mesoporous structure, enlarged surface area and multiple scattering enable it to harvest light much more efficiently.(Yu, Wang et al. 2004) This enhanced light-trapping effect is the result of the reflection or transmission of the light scattered by the macroporous tunnels or mesopores implanted in the body of Al, Fe co-doped TiO_2 matrix. It is also noted that the modification of Al_2O_3 did not change the light sensitization region. However, a very weak blue shift to short wavelength was observed for Al-Fe/TiO_2sample after 700° C calcining treatment. The band energy gap of the as-prepared samples could be calculated by using $(\alpha h\nu)^n = k(h\nu - E_g)$, where α is the absorption coefficient, k is the parameter that related to the effective masses associated with the valence and conduction bands, n is 1/2 for a direct transition, $h\nu$ is the absorption energy, and E_g is the band gap energy.(Li, Zhang et al. 2009) Plotting $(\alpha h\nu)^{1/2}$ versus $h\nu$ based on the spectral response in Figure 4a gave the extrapolated intercept corresponding to the E_g value (see Figure 4b). The optical band energies of the macro/mesoporous TiO_2-400, Fe/TiO_2-400, Al-Fe/TiO_2-400 and Al-Fe/TiO2-700 samples (3.10 eV, 2.50 eV, 2.60 eV, and 2.66 eV respectively) exhibit obvious red-shifts with respect to that of TiO_2-400 sample (3.10 eV). The results of this study therefore indicate that the enhanced ability to absorb visible-light of this type of macro/mesoporous Al-Fe/TiO_2 makes it a promising photocatalyst for solar-driven applications.

To evaluate the photocatalytic performance of the as-prepared materials. The photo-oxidation of NO gas under visible light irradiation (λ > 400 nm) in a single pass flow was used as a photoreaction probe.Figure 5a shows the relative of NO removal rate against irradiation time in the presence of photocatalysts under visible-light irradiation. In the absence of the photocatalyst, no obvious removal rate of NO can be observed. The Photocatalytic performance of pure TiO_2 can be nearly neglected. The NO removal rate over Fe/TiO_2-400 sample reaches 17 % after 20 min irradiation, indicating the promotion effect of Fe-doping. Compared to Fe/TiO_2-400, Al-Fe/TiO_2-400 exhibits a much higher removal rate (about 28 % after 20 min irradiation). Such high photocatalytic performance maybe attributed to the high surface area, large pore volume. Besides, 3D connected pore tunnels are also very important because they can allow the NO molecule to transport very conveniently in the body of the catalyst. Further increasing the calcinations temperature to 700° C quickly decreased the removal rate to about 10 %.

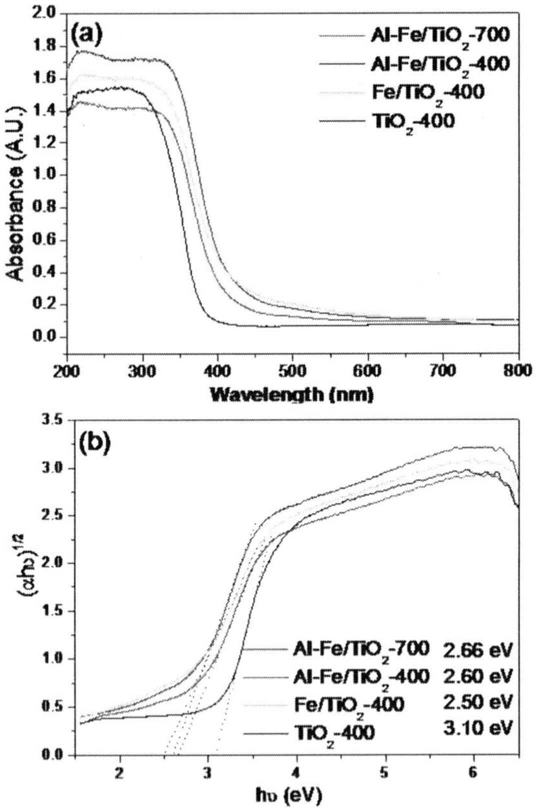

Figure 4. UV-visible absorption spectra of (a) and determination of indirect interband transition energies (b) for pure TiO_2, Fe/TiO_2-400, Al-Fe/TiO_2-400, and Al-Fe/TiO_2-700 samples.

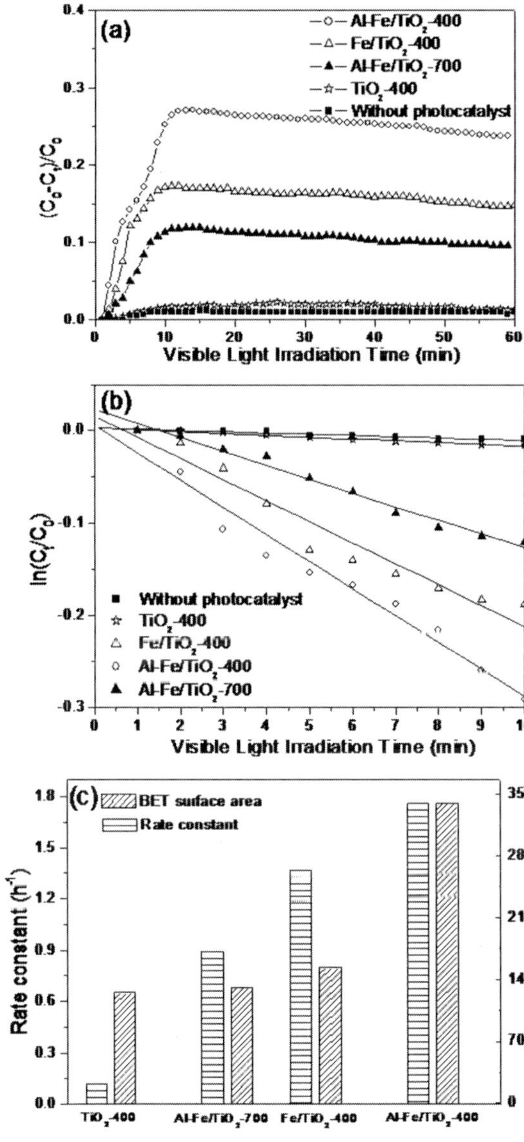

Figure 5. a) Plots of the removal of NO concentration vs irradiation time in the presence of the as-prepared products with visible-light irradiation (> 400 nm). (b) Dependence of ln(C/C$_0$) on irradiation time. (c) Relationship between rate constant and BET surface area over the as-prepared products.

For a clear quantitative comparison, we use the Langmuir-Hinshelwood model (L-H) to describe the initial rates of photocatalytic removal of NO. The photocatalytic oxidation of NO was recognized to follow a first-order-kinetics approximately as a result of low concentration target pollutants, as evidenced by the linear plot of ln (C/C_0) versus photocatalytic reaction time t (Figure 5b). The rate constants of the TiO_2-400, Fe/TiO_2-400, Al-Fe/TiO_2-400 and Al-Fe/TiO_2-700 samples are 0.119 h^{-1}, 1.368 h^{-1}, 1.762 h^{-1} and 0.893 h^{-1} respectively. Figure 5c shows the relationship between reaction rate constants and BET surface areas. Al-Fe/TiO_2-400 sample owns the highest surface area of 340 m^2/g, resulting in an excellent photocatalytic performance in oxidation of NO. Though the surface area of Al-Fe/TiO_2-700 is similar to that of Fe/TiO_2-400, the reaction rate constant of the formed is much lower that of the latter. Except for the effect of surface area, other factors such as the band gap and light adsorption capability also play an important role in controlling the photocatalytic performance of the catalysts. As shown in Figure 4b, the band gap of Al-Fe/TiO_2-700 is 2.66 eV is higher than that of Fe/TiO_2-400. Meanwhile, the light adsorption intensity of Al-Fe/TiO_2-700 is much lower than that of Fe/TiO_2-400.

4. CONCLUSIONS

Macro/mesoporous Fe/TiO_2 was fabricated by soft-chemical synthesis in the presence of surfactants, followed by calcination. Such materials have been proved as a good photocatalyst for treating NO at air conditions for its special macro/mesoporous structures. The modification of Al_2O_3 can effectively increase the thermal stability of Fe/TiO_2 with a very high surface area, resulting in an excellent photocatalytic performance during the oxidation of 400 ppb level of NO in air under visible light irradiation. The present work demonstrates that the hierarchical macro/mesoporous Fe/TiO_2photocatalysts are effective visible-light-driven photocatalytic functional materials for air purification.

5. ACKNOWLEDGEMENTS

This work was supported by the Program for Professor of Special Appointment (Eastern Scholar) at Shanghai Institutions of Higher Learning, the National Natural Science Foundation of China (21007040, 21047009), Natural Science Funding of Shanghai (11ZR1426300), the Research Fund for the Doctoral Program of Higher Education (20103127120005), the Project supported by the Shanghai Committee of Science and Technology (10160503200), and by a Scheme administrated by Shanghai Normal University (8K201104).

REFERENCES

1 X. Chen, X. Wang, X. Fu, 2009 Hierarchical Macro/Mesoporous TiO2/SiO2 and TiO2/ZrO2 Nanocomposites for Environmental Photocatalysis. Energy & Environmental Science, 2 8 (May 2009), 872877 , ISSN: 1754-5692

2 G. Li, D. Zhang, J. Yu, 2009 Thermally Stable Ordered Mesoporous CeO2/TiO2 Visible-Light Photocatalysts. Physical Chemistry Chemical Physics, 11 19 (February 2009), 37753782 , 1463-9076

3 M. Machida, K. Norimoto, T. Watanabe, K. Hashimoto, A. Fujishima, 1999 The Effect of SiO2 Addition in Super-Hydrophilic Property of TiO2 Photocatalyst. Journal of Materials Science, 34 11 (June 1999), 25692574 . 0022-2461

4 S. Nahar, K. Hasegawa, S. Kagaya, 2006 Photocatalytic Degradation of Phenol by Visible Light-Responsive Iron-Doped TiO2 and Spontaneous Sedimentation of the TiO2 Particles. Chemosphere, 65 11 (December 2006), 19761982 . 0045-6535

5 D. Rolison, 2003 Catalytic Nanoarchitectures- The Importance of Nothing and the Unimportance of Periodicity. Science, 299 5613 (May 2003), 16981701 . 0036-8075

6 X. Wang, J. Yu, Y. Chen, L. Wu, X. Fu, 2006 ZrO2-Modified Mesoporous Manocrystalline TiO2-xNx as Efficient Visible Light Photocatalysts. Environmental Science & Technology, 40 7 (April 2006), 23692374 , 0001-3936X

7 X. Wang, J. Yu, C. Ho, Y. Hou, X. Fu, 2005 Photocatalytic Activity of a Hierarchically Macro/Mesoporous Titania. Langmuir, 21 6 (March 2005), 25522559 , 0743-7463

8 N. Wu, S. Wang, I. Rusakova, 1999 Inhibition of Crystallite Growth in the Sol-Gel Synthesis of Nanocrystalline Metal Oxides. Science, 285 5432 (August 1999), 13751377 , 0036-8075

9 J. Yu, G. Li, X. Wang, X. Hu, C. Leung, Z. Zhang, 2006 An Ordered Cubic Im3m Mesoporous Cr-TiO2 Visible Light Photocatalyst. Chemical Communications, 25 (September 2006), 27172719 , 1359-7345

10 J. C. Yu, X. Wang, X. Fu, 2004 Pore-Wall Chemistry and Photocatalytic Activity of Mesoporous Titania Molecular Sieve Films. Chemistry of Materials, 16 8 (April 2004), 15231530 , 0897-4756

11 Z. . Yuan, L. Su, 2006 Insights into Hierarchically Meso-Macroporous Structured Materials. Journal of Materials Chemistry, 16 7 (Febrary2006), 663677 , 0959-9428

12 J. Zhu, J. Ren, Y. Huo, Z. Bian, H. Li, 2007 Nanocrystalline Fe/TiO2 Visible Photocatalyst with a Mesoporous Structure Prepared via a Nonhydrolytic Sol-Gel Route. Journal of Physical Chemistry C, 111 51 (December 2007), 1896518969 , 1932-7447

CHAPTER 3

Evaluation of Some Radioactive Materials and Heavy Metals in Marine Environment of Alexandria Coastline, Egypt

E. R. Atta*, Kh. M. Zakaria

Egyptian Nuclear and Radiological Regulatory Authority, Cairo, Egypt

ABSTRACT

The present work is carried out to obtain quantitative information of some natural radioactive materials and heavy metals in water and sediment samples along the Alexandria Coastline, as a step to construct the baseline map of the background radioactivity level in the Egyptian environment and also as a base data to assess the future physicochemical changes of surface coastal water and sediment in the studied area. The distribution of natural gamma emitting radionuclides such as ^{238}U, ^{232}Th and ^{40}K has been established by gamma spectrometry.

The results show that the area of Alexandria coast is affected by the existence of high back-ground radiation from the Rashid coast at the eastern side which has high levels of radioisotope concentrations due to black sand deposits which are dominant in this area. The absorbed dose rate (D, nGy·h^{-1}), annual effective dose equivalent, external hazard index (H_{ex}) and representative level index (I_r) of the investigated radioactive materials were calculated. The concentration of Pb^{2+}, Cd^{2+}, Ni^{2+}, Mn^{2+}, Cu^{2+}, Co^{2+}, Zn^{2+} and Cr^{3+} ions has been determined using Atomic Absorption Spectroscopy. The total dissolved solids (TDS) in water samples ranged from 33,000 mg/l to 42,000 mg/l, the salinity ranged from 37.9% to 40.5% and pH ranged from 7.6 to 7.9. Some of the sediment quality guidelines are calculated and they are the metal pollution index (MPI), the contamination factor (CF), degree of contamination (C_{deg}), and Pollution load index (PLI).

KEYWORDS

Mediterranean Sea, Water, Sediment, Radioactive Materials, Heavy Metals, Sediment Quality Guidelines

1. INTRODUCTION

Pollution is the most serious of all environmental problems and posses a major threat to the health and well- being of millions of people and global ecosystems. In recent years, pollution becomes a paramount problem with increasing the human activities. Natural and artificial radionuclides have attained great attention of many researchers in the last three decades especially after Chernobyl radiation accident in 1986. Many authors investigated the Ocean waters [1] and marine sediments [2] - [4] . The sources of radiation in water are also the sources in the case of sediment. The most dominant radionuclides are ^{40}K and also decay products of the ^{238}U and ^{232}Th series [5] . Most of radionuclides have a low solubility in water and tend to be sorbed onto the particulate matter, therefore they accumulate in sediments. Fine sediments, with their large surface area, tend to sorb more than coarse sediments. Thus, sea water has a radioactivity of about 12.6 $Bq \cdot kg^{-1}$, while marine sand has a radioactivity of 200 - 400 $Bq \cdot kg^{-1}$, and mud 700 - 1000 $Bq \cdot kg^{-1}$, in parts of the world [5] . The pollution of marine ecosystems by heavy metals is a world-wide problem. Generally, the coastal industries appear to discharge wastes directly into the seas environments with little or no treatment. Mediterranean Sea is a semi-closed basin connected to the Atlantic Ocean through the narrow and shallow Strait of Gibraltar [6] . So the ecosystem of the Mediterranean Sea affected by chemical pollutant resulted from the human activities. The contamination levels of the aquatic environment by heavy metals can be estimated by analyzing its water and sediments. Heavy metals are considered a major contaminant in coastal and marine environments worldwide [7] . They pose a serious threat to human health, living organisms and natural ecosystems because of their toxicity, persistence and bioaccumulation characteristics [8] . Many heavy metal ions are known to be toxic or carcinogenic to human [9] . Metals such as copper, zinc and manganese are essential metals since they play important roles in biological systems [10] , whereas non-essential metals such as Pb, and Cd are usually potent toxins and their bioaccumulation in tissues leads to intoxication, decreased fertility, cellular and tissue damage, cell death and dysfunction of a variety of organs [11] [12] . Heavy metals can contribute in degradation of marine ecosystems by reducing species diversity and through accumulation of metals in living organisms [13] .

The coastal states of the Mediterranean Sea do their efforts to protect the nature of the sea. Egypt has about 1050 km of coastline along the Mediterranean Sea proper, which is of great environmental, economical and recreational value. Some of the Egyptian coastal areas of the Mediterranean Sea (especially in front of the large cities) receive different types of pollution sources, so the study area located between 31°08'11"N into 31°21'18"N and 29°50'02"E into 30°18'20"E. The present work is carried out to obtain quantitative information on some natural radioactive materials and heavy metals in water and sediment samples along the Alexandria Coastline, as a step to construct the baseline map of the background radioactivity level in the Egyptian environment and also as a base data to assess the future physicochemical changes of coastal surface water and sediment in the study area.

2. MATERIALS AND METHODS

2.1. Study Area

The investigated area (110 km along) expanded along the coast of Alexandria, from Al-Max 31°08'11"N to 29°5'02"E in west to Abu-Qir 31°17'20"N to 30°08'23"E in east and stretches to Rashid coast city 31°21'18"N to 30°18'20"E.

2.2. Sampling and Sample Preparation

Collected samples in the coastline study area occurred during spring 2012. Five representative surficial shore sediment samples were collected using the template method at the same sites and time of seawater samples. An area of about 25 ´ 25 cm^2 up to a depth of 5 cm was cut out using the stainless steel template for guidance [14] . The collected shore sediment samples were transferred to labelled polyethylene bags, closed and transported to the laboratory for preparation and chemical analysis.

The shore sediment samples were air-dried at room temperature for a week. And also were dried in an oven at 80°C (for 48 h) till constant dry weight was obtained, crushed and homogenized. Then milled and sieved through 0.4 mm mesh sieves and stored for further analysis. Water samples, 5 liter of each, were collected using the water sampler. They were collected in polyethylene containers. Then, the samples were acidified with Nitric acid to pH lower than 2 to avoid micro-organisms growth and to minimize water-walls interaction. The samples were stored for radioactivity measurements and chemical analysis [14] . The total dissolved solid (TDS), salinity, pH and temperature (°C) were measured in the field after sampling. These measurements carried out according to [15] .

2.3. Heavy Metals Measurements in Sediment and Water Samples

Atomic absorption Spectrophotometer (AAS) is a simple and well available technique for the determinations of heavy metals in the water and soil samples. Heavy metals in sediments were determined according to [16] . The sediments were digested with 5:1 mixture of HF and $HClO_4$acids. 1 g (dry weight) sample was digested by 2 ml $HClO_4$ and 10 ml HF to near dryness, subsequently a second addition of 1 ml of $HClO_4$ and 10 ml of HF and evaporated to near dryness. Finally, 1 ml of $HClO_4$ alone was added and the sample was evaporated until the appearance of white fumes. The residue was dissolved in 12 N HCl and diluted to 25 ml with de-ionized water. The metal ions were determined by Atomic Absorption Spectrophotometer, Perkin Elmer model Aanalyst 100 which is manufactured in USA. The results obtained were determined according to [17] .

2.4. Radioactivity Measurements for Sediment and Water Samples

2.4.1 Radioactivity Measurements for Sediment Samples

1) Sample preparation

The homogenized sediment samples were packed in a 250 ml plastic container to its full volume with uniform mass. These containers were shielded to ensure that all daughter products of Uranium and Thorium, in particular radon gas formed, do not escape. The net weight of the sample was determined before counting. These samples were then stored for 30 - 40 days before counting to reach radioactive equilibrium.

2) Radioactivity measurements

The activity concentration of the natural radioactivity ^{238}U, ^{232}Th and ^{40}K in the investigated samples were determined using a high-resolution HPGe γ-spectrometry system with 30% counting efficiency. This was performed by taking 250 cm^3 counting vials filled up to a height of 7 cm, which correspond to 170 cm^3. The measurement duration was up to 80,000 sec and were carried out in the Laboratory of Egyptian Nuclear and Radiological Regulatory Authority. The obtained spectra were analyzed. The determination of the presence of radionuclides and calculation of their activities were based on the following gamma-ray transitions (in keV): the ^{226}Ra activities (or ^{238}U activities for samples assumed to be in radioactive equilibrium) were estimated from ^{234}Th (92.38 keV, 5.6%), while γ-energies of ^{214}Pb (351.9 keV, 35.8%) and ^{214}Bi (609.3, 45%), 1764.5 keV, 17%) and ^{226}Ra (185.99 KeV, 3.5%) were used to estimate the concentration of ^{226}Ra. The Gamma-ray energies of ^{212}Pb (238.6 keV, 45%), and ^{228}Ac (338.4 keV, 12.3%), (911.07 keV, 29%), (968.90 keV, 17%) were used to estimate the concentration of ^{232}Th. The activity concentrations of ^{40}K were measured directly by its own gamma rays (1460.8 keV, 10.7%). In order to determine the background distribution due to naturally occurring radionuclides in the environment around the detector, an empty polystyrene container was counted in the same manner as the samples. The activity concentrations were calculated after measurement and subtraction of the background. The activities were determined from measuring their respective decay daughters. The activity concentrations were calculated from the intensity of each line taking into account the mass of the sample, the branching ratios of the γ-decay, the time of counting and the efficiencies of the detector [18] [19] . The activity concentrations of the investigated samples were calculated from Equation (1):

$$A = \frac{(CPS)_{net}}{I \cdot E_{ff} \cdot M} \qquad (1)$$

where A is the activity concentration in Bq/kg, $(CPS)_{net}$ is the (count per second) and equals

$\left[(cps)_{sample} - (cps)_{background}\right]$, I is the intensity of the γ-line in a radionuclide, E_{ff} is the measured efficiency for

each γ-line observed and M is the mass of the sample in kilograms. The correction for the contribution of ^{232}Th via its daughter nuclide ^{228}Ac (1459.2 keV peak) to the 1460.8 keV peak of ^{40}K was made according to following formula [20] .

$$\text{The error in } ^{40}\text{K activity } (\%) = 9.3\left(A_{Th}/A_K\right) \qquad (2)$$

where: A_{Th} and A_K are the activity concentration of ^{232}Th and ^{40}K, respectively, in $Bq{\cdot}kg^{-1}$.

3. RESULTS AND DISCUSSION

Physical characteristics of water samples at different locations of studied area Table 1 represents some physical characteristics of water samples at different locations of studied area. It is clear that the total dissolved solids (TDS) in water samples ranged from 33,000 mg/l to 42,000 mg/l, the salinity is fluctuated pattern and ranged from 37.9% to 40.5% with an average of 39.02% and the marine water is slightly alkaline (pH ranged from 7.6 to 7.9).

3.1. Heavy Metals in Water

The results corresponding to the seawater analysis of the different samples are shown in Table 2and Figure 1. In general, the analyzed heavy metals showed increasing concentrations of Mn and Zn in all studied locations. It is noted that Cd concentration is the lowest in all locations, while the concentration of Pb, Ni, Cr, Cu and Co ions are in the same level. In general, this may be due to anthropogenic sources, a mixture of contaminated sediments with relatively clean marine sediments and/or the release of metals into the water as a freshwater and seawater combination [21] .

3.2. Heavy Metals in Sediment

The concentrations of heavy metals in the investigated sediment samples of Alexandria coast are shown in Table 3 and Figure 2. It is clear that the most abundant contaminated ions in all locations are Mn, and Zn, Mn > Zn. while Cd is the lowest contaminant ion concentration for all locations. Eastern Harbour has highest concentration of Cr ions; this may be attributed to the effect of ships discharges which used antifouling paints.

El-Dekhila has highest concentrations of both Co and Pb; this may be due to the effect of El-Dekhila electric power station which use metal alloys in containers for heating the water steam to produce electricity. Rashid and El-Mex locations have higher concentration of Cu ions. The variety of the concentrations of heavy metals in bulk sediments of Alexandria coast may be attributed to the effect of type and amount of pollutant arrives to it and the characteristic nature of the area [22] .

3.3. Sediment Quality Guidelines

The studied parameters of sediment quality guidelines are the metal pollution index (MPI), the contamination factor (CF), degree of contamination (C_{deg}), and Pollution load index (PLI).

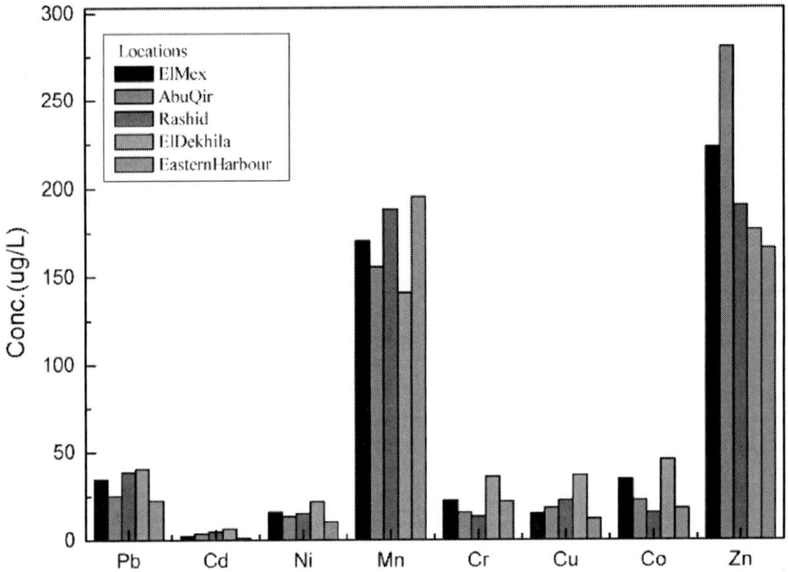

Figure 1. Concentration of some heavy metal ions in marine water at different locations along the study area.

Table 1. Some physical characteristics of water samples at different locations of studied area.

Location	pH	TDS mg/l	Salinity (%)	Temp. (°C)
El-Mex	7.8	40,000	38.7%	20.1
Eastern Harbor	7.6	42,000	38.6%	22.7
El-Dekhila	7.8	41,000	40.5%	20.9
Abu-Qir	7.7	36,000	37.9%	22.4
Rashid	7.9	33,000	39.4%	21.5

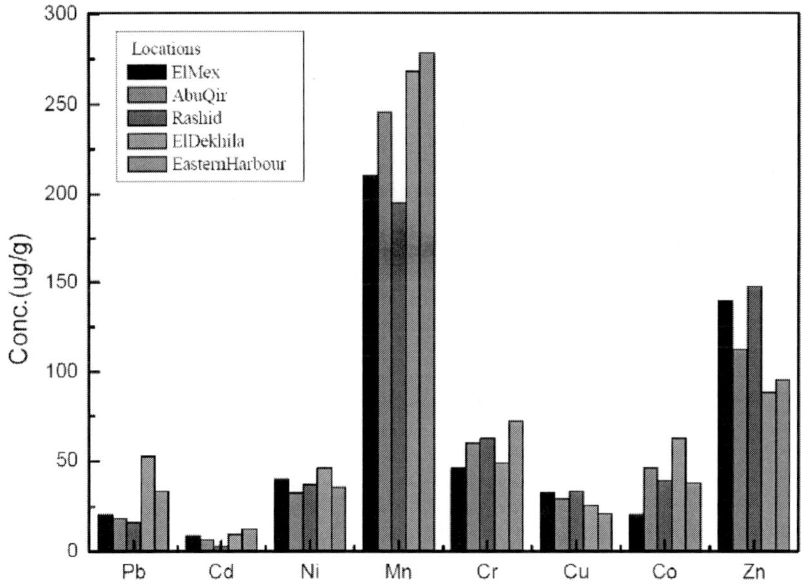

Figure 2. Concentration of some heavy metal ions in sediment samples at different locations in study area.

Table 2. Concentration of some heavy metals in marine samples.

Elements (μg/g)	Locations and their coordinates				
	El-Mex 31°08'11"N 29°50'02"E	Eastern Harbour 31°12'16"N 29°53'16"E	El-Dekhila 31°09'07"N 29°48'09"E	Abu-Qir 31°17'20"N 30°08'23"E	Rashid 31°21'18"N 30°18'20"E
Pb	34.6 ± 7.2	22.7 ± 4.6	40.5 ± 5.1	25.3 ± 4.1	38.7 ± 4.8
Cd	2.4 ± 0.2	1.4 ± 0.05	6.7 ± 1.4	3.8 ± 0.9	4.8 ± 1.1
Ni	16.2 ± 2.3	10.5 ± 1.2	21.8 ± 3.2	13.7 ± 2.1	15.4 ± 1.9
Mn	170.4 ± 11.1	195.2 ± 15.7	140.7 ± 8.2	155.6 ± 9.2	188.4 ± 12.8
Cr	22.7 ± 2.1	21.8 ± 1.9	35.9 ± 4.6	15.9 ± 1.8	13.5 ± 1.2
Cu	15.3 ± 2.3	12.3 ± 1.6	36.8 ± 4.8	18.4 ± 3.1	22.7 ± 4.2
Co	34.6 ± 7.2	18.2 ± 1.9	45.6 ± 8.5	22.8 ± 2.8	15.7 ± 2.3
Zn	223.5 ± 13.8	166.3 ± 8.7	176.8 ± 9.5	280.7 ± 15.6	190.4 ± 11.6

Table 3. Concentration of some heavy metals in sediment samples and the metal pollution index (MPI).

Elements (μg/g)	Location					Residential soil quality (μg/g)
	El-Mex	Eastern Harbour	El-Dekhila	Abu-Qir	Rashid	
Pb	20.45 ± 0.44	33.62 ± 1.28	52.74 ± 4.6	18.3 ± 0.28	15.95 ± 0.21	140.0*
Cd	8.68 ± 0.34	12.55 ± 0.48	9.77 ± 0.42	6.42 ± 0.27	2.82 ± 0.13	10.0*
Ni	40.54 ± 3.20	35.77 ± 1.33	45.96 ± 3.1	32.85 ± 1.58	37.52 ± 1.2	50.0*
Mn	210.33 ± 9.55	278.46 ± 8.2	268.22 ± 11.3	245.88 ± 12.9	195.28 ± 12.8	1100**
Cr	45.77 ± 4.66	72.33 ± 5.9	48.63 ± 3.7	59.86 ± 5.2	62.48 ± 4.8	64.0*
Cu	32.66 ± 1.12	20.87 ± 0.67	25.87 ± 1.35	29.57 ± 0.87	33.85 ± 1.23	63.0*
Co	20.66 ± 4.6	38.33 ± 2.6	62.53 ± 4.25	45.95 ± 5.8	39.76 ± 3.24	50.0*
Zn	139.22 ± 4.55	95.25 ± 3.55	88.35 ± 2.2	112.44 ± 2.88	147.62 ± 5.2	200.0*
MPI	39.96	29.69	50.77	41.43	37.71	

*Canadian soil quality guidelines (2002) for the protection of environmental and human health; **Persaud *et al.* (1993).

3.4. The Metal Pollution Index (MPI)

The degree of heavy metal pollution can be estimated using the metal pollution index (MPI) according to the following formula:

$$MPI = \left(Pb \times Cd \times Ni \times Mn \times Cr \times Cu \times Co \times Zn\right)^{1/8} \quad (4)$$

The concentration of a metal is expressed in $\mu g \cdot g^{-1}$; dry weight [23] .

According to the calculated data of the metal pollution index (MPI), (Table 3), it can be concluded that the highest values (50.77 and 41.43) were found at El Dekhela and Abu-Qir locations, respectively. The lowest (29.69) value were found at Eastern Harbour.

3.5. The Contamination Factor (CF)

Contamination factor is an assessment of soil contamination through a reference element in comparison with crustal level. It is an effective tool for monitoring pollution over a period of time and evaluating the pollution of environmental single substances. It is defined according to four categories as follows [24] .

(CF < 1 low contamination factor indicating low contamination; 1 < CF < 3 moderate contamination factor; 3 < CF < 6 considerable contamination factor; CF > 6 very high contamination factor).

Individual contamination factors are calculated based on the following formula:

$$CF = M_x / M_b \tag{5}$$

where: M_x is the concentration of the target metal and M_b is the concentration of the metal in the selected reference background.

The results of contamination factors (Table 4) indicated that Cd and Cr possess the highest CFs in Eastern Harbour, also El-Dekhila has highest CF of Co. They reflect that sediments in these locations are moderately contaminated (1 < CF < 3). The rest of metals in all locations are remained within the low contamination level of the sediment (CF < 1).

Table 4. Contamination factor (CF), degree of contamination (deg. cont.) and pollution load index (PLI) of some heavy metals in soil sediment samples at studied locations.

CF of elements	Location				
	El-Mex	Eastern Harbour	El-Dekhila	Abu-Qir	Rashid
CF of Pb	0.16	0.24	0.38	0.13	0.11
CF of Cd	0.87	1.26	0.98	0.64	0.28
CF of Ni	0.81	0.72	0.92	0.66	0.75
CF of Mn	0.19	0.25	0.24	0.22	0.18
CF of Cr	0.72	1.13	0.76	0.94	0.98
CF of Cu	0.52	0.33	0.41	0.47	0.54
CF of Co	0.41	0.77	1.25	0.92	0.80
CF of Zn	0.70	0.48	0.44	0.56	0.74
Deg. cont.	4.38	5.18	5.38	4.54	4.38
PLI	0.47	0.54	0.59	0.48	0.44

3.6. Degree of Contamination (C_{deg})

The contamination factor described above is a single element index. The sum of contamination factors for all elements examined represents the contamination degree (C_{deg}) of the environment [25] reported that there are four classes of contamination degree. They are $C_{deg} < 8$ low degree of contamination, $8 < C_{deg} < 16$ moderate degree of contamination, $16 < C_{deg} < 32$ considerable degree of contamination, and $32 < C_{deg}$ very high degree of contamination.

The sums of contamination factors for all metals examined are listed in Table 4 which shows that the degree of contamination in the investigated sediment samples ranged from 4.38 to 5.38. This indicates that the degree of contamination in the investigated sediment samples in all locations is low. Also it can be seen from the data that the principle contributors among toxic trace metals are Zn, Mn, Ni and Cu ions.

3.7. Pollution Load Index (PLI)

The pollution load index (PLI) of each site was evaluated as indicated by Tomilson et al. (1980) [26] .

$$\text{Pollution load index} = \left(CF_1 \times CF_2 \times CF_n\right)^{1/n} \tag{6}$$

where; n is the number of metals and CF is the contamination factor. Contamination can be estimated as follows: (PLI value > 1 polluted; PLI value < 1 unpolluted [23] [27]). The PLI values calculated for each site were shown in Table 3. It is shown that the PLI for all locations ranged from 0.44 to 0.59. This indicates that all studied sites were found to be low polluted (PLI < 1).

3.8. Activity of the Investigated Radionuclides in Water and Sediment Samples

Table 5 and Figures 3-6 represent the activity of ^{238}U, ^{232}Th and ^{40}K in coastal seawater and sediment samples at different locations along the studied area. It is clear that the measured values for ^{238}U and ^{232}Th in sea water are very low. The ^{238}U values vary from 13.08 Bq·L^{-1} at El Mex location to 17.30 Bq·L^{-1} at Rashid location. The highest value at Rashid could be attributed to the existence of black sand. Also, the ^{232}Th values are low and varies from 10.32 Bq·L^{-1} at Eastern Harbour location to 20.33 at Bq·L^{-1} Abu-Qir location. The distribution of ^{40}K in sea water along the locations revealed that, the Al-Mex is poor in ^{40}K (562.04 Bq·L^{-1}) while Abu-Qir location has a highest activity (1273.02 Bq·L^{-1}) and it has differences in their patterns. It is clear that the lower ^{40}K has been detected in locations receiving low saline water discharged from inland sources such as Rashid branch and outlets of coastal lakes. Because ^{40}K is an alkali metal, it behaves as a true soluble element and varies in proportion with respect to salinity. So the distribution of ^{40}K in surface seawater and in sediments has differences in their patterns due to the effect of salinity on the behavior of ^{40}K in surface seawater. It is observed that the general trend of natural radionuclides in the sediments decreases in the direction of east locations to the west locations, Figure 6. Also it is noticed that large variation among the radioactivity concentration for different locations. It may be due to geological condition and drainage pattern of the study area.

It is clear that there is a general decreasing trend of ^{238}U from Rashid and Abu-Qir to Al-Max location. This is because the sea wave direction in this region is mostly to the west, which helps in the deposition of the eroded black sand from the Rashid coast which is about 30 km East Abu-Qir.

The ^{232}Th measured values in sediment along the study area more or less have the same trend as^{238}U. The high values could be explained as due to the presence of black sands, which are enriched in the mineral monazite containing a significant amount of ^{232}Th. The black sands, which is rich in radio-nuclides, supplied by the River Nile during the flooding period in the last periods through the Rashid branch of the Nile. The Mediterranean Sea waves carry part of these black sands to Alexandria beaches westward.

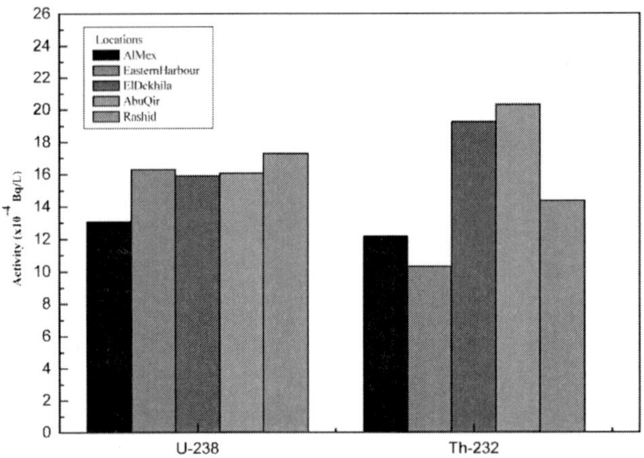

Figure 3. Activity of ^{238}U and ^{232}Th in marine water at different locations along the study area.

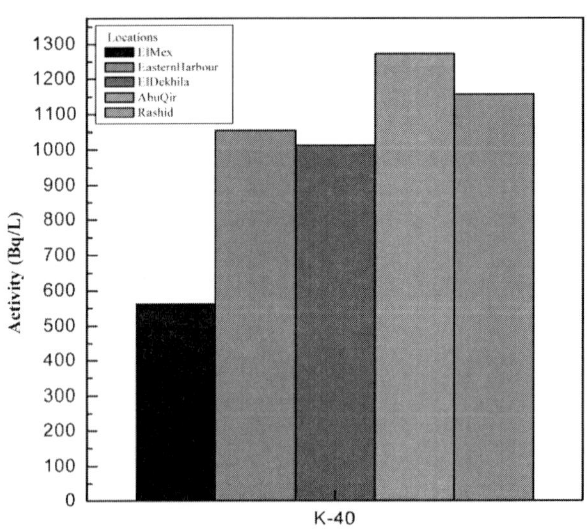

Figure 4. Activity of ^{40}K in marine water at different locations along the study area.

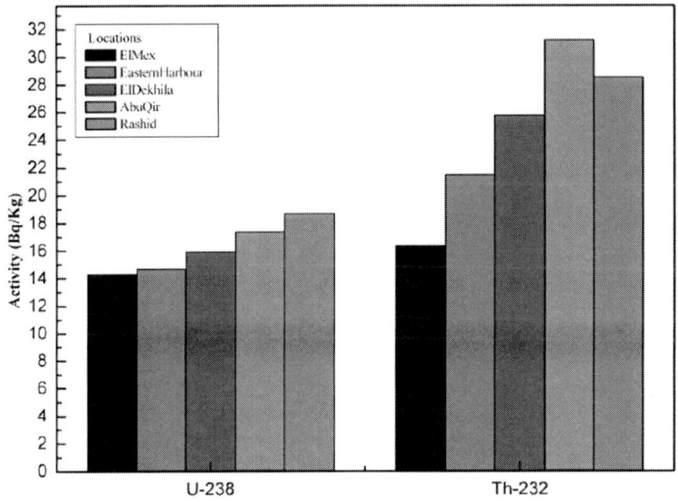

Figure 5. Activity of ^{238}U and ^{232}Th in sediment samples at different locations along the study area.

Table 5. Concentration of ^{238}U, ^{232}Th and ^{40}K in seawater and coastal sediments collected from different locations along the studied area.

Radionuclides	Locations									
	El-Mex		Eastern Harbour		El-Dekhila		Abu-Qir		Rashid	
	Water (Bq/L)	Sediment (Bq/kg)	Water (Bq/L)	Sediment (Bq/kg)	Water (Bq/L)	Sediment (Bq/kg)	Water (Bq/L)	Sediment (Bq/kg)	Water (Bq/L)	Sediment (Bq/kg)
^{238}U	13.08 ± 0.04	14.28 ± 0.55	16.32 ± 0.41	14.69 ± 0.58	15.90 ± 0.33	15.90 ± 0.65	16.08 ± 0.18	17.38 ± 0.34	17.30 ± 0.24	18.09 ± 0.25
^{232}Th	12.18 ± 0.25	16.34 ± 0.88	10.32 ± 0.28	21.46 ± 0.55	19.27 ± 0.34	25.73 ± 0.65	20.33 ± 0.31	31.21 ± 0.92	14.37 ± 0.22	28.52 ± 0.71
^{40}K	562.04 ± 3.5	644.67 ± 1.22	1053.42 ± 6.4	422.20 ± 0.43	1013.20 ± 0.98	567.94 ± 0.99	1273.0 ± 0.83	583.79 ± 1.13	1156.7 ± 1.23	988.20 ± 2.86

3.9. Determination of the Absorbed Dose Rate, Annual Effective Dose Equivalent, External Hazard Index and Representative Level Index Values

The absorbed dose rate was calculated from the measured activities of ^{238}U, ^{232}Th and ^{40}K in the surface sediment samples using the below formula [28] .

$$D\left(nGy \cdot h^{-1}\right) = 0.462C_U + 0.604C_{Th} + 0.042C_K \quad (7)$$

where: D is the absorbed dose rate ($nGy \cdot h^{-1}$), C_U, C_{Th} and C_K are the activity concentrations ($Bq \cdot kg^{-1}$) of ^{238}U, ^{232}Th and ^{40}K respectively. To estimate the annual effective dose rates, the conversion coefficient from absorbed dose to effective dose, 0.7 $Sv \cdot Gy^{-1}$ and outdoor occupancy factor of 0.2 proposed by UNSCEAR, 2000 were used. The effective dose rate in units of $mSv \cdot y^{-1}$was calculated by the following formula:

$$\text{Effective dose rate}\left(mSv \cdot y^{-1}\right) = D\left(nGy \cdot h^{-1}\right) \times 8760\ h \times 0.2 \times 0.7\ Sv \cdot Gy^{-1} \times 10^{-6} \quad (8)$$

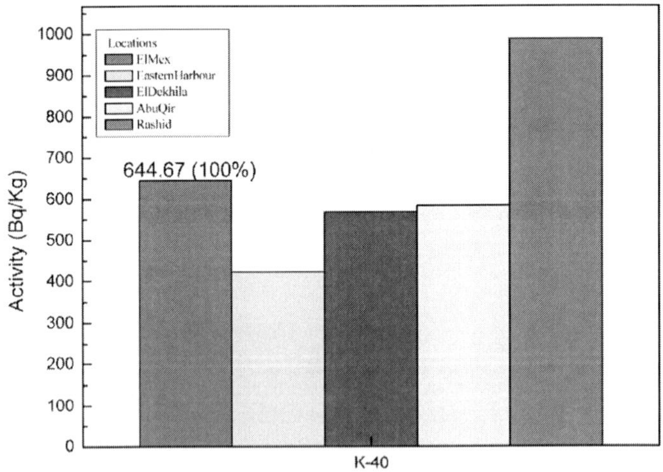

Figure 6. Activity of ^{40}K in sediment samples at different locations along the study area.

Calculation of hazard indexes (H_{ex}):

The external hazard index, H_{ex}, was calculated using the following formula [29] .

$$H_{ex} = \frac{C_{^{238}U}}{370} + \frac{C_{^{232}Th}}{259} + \frac{C_{^{40}K}}{4810} \leq 1 \tag{9}$$

An additional hazard index so called representative level index is calculated by using the following formula [23] ,

$$I_r = \frac{C_{^{238}U}}{150} + \frac{C_{^{232}Th}}{100} + \frac{C_{^{40}K}}{1500} \tag{10}$$

The value of these indexes must be less than unity in order to keep the radiation hazard insignificant.

Table 6 represents the calculated absorbed gamma dose rate from 33.73 $nGy \cdot h^{-1}$ (El-Max) to 67.46 $nGy \cdot h^{-1}$ (Rashid), with a mean of 47.02 $nGy \cdot h^{-1}$. The mean absorbed dose rate is found to be lower than the world average value (51 $nGy \cdot h^{-1}$: UNSCEAR, 2000). The calculated values of annual effective dose rate ranging from 0.041 to 0.083 mSv, with a mean value of 0.058 mSv, which is lower than the world average value of 0.48 mSv (UNSCEAR, 2000). The calculated value of external hazard index ranges from 0.207 to 0.366. The representative level index value being 0.577 to 0.978, with the average of 0.762, which is higher than the world average (0.66 $Bq \cdot kg^{-1}$) [30] .

4. CONCLUSIONS

The results demonstrate that the main source of radiation along the Alexandria coast is the black sand supplied by the River Nile through the Rashid branch. There is a general decreasing trend of ^{238}U and ^{232}Th from Rashid and Abu-Qir to Al-Max westerly. The distribution of ^{40}K along the Alexandria coast reveals that the Al-Max location is poor in ^{40}K, while Abu-Qir is rich in ^{40}K. There is no pattern for the concentration of ^{40}K in sediments along the studied locations.

The mean absorbed dose rate is found to be lower than the world average value (51 $nGy \cdot h^{-1}$). The mean annual effective dose rate is lower than the world average value (0.48 mSv). The average external hazard index value is higher than the world average (0.66 $Bq \cdot kg^{-1}$).

Spatial distribution of heavy metals in sediment of the study area revealed that Alexandria coast may be affected by different sources of pollution (sewage and garbage from the Alexandria City and ships awaiting transit area, industrial effluents including oil refineries, fertilizer plant, power stations and other industries).

The index of contamination factor shows that the sediment samples from eastern harbour location are moderately contaminated with chromium and El-Dekhila location is moderately contaminated with cobalt. All other metals analyzed (Pb, Cd, Ni, Mn, Cu and Zn) display low level of contamination in sediment samples. The sum of contamination factors for all metals examined indicates low degree of contamination in sediment. The degree of contamination (C_{deg}) in sediment samples is low in all locations.

Table 6. External hazard index (H_{ex}) values, representative level index (I_r) values, absorbed dose rate and annual effective dose rate for radioactive materials (^{238}U, ^{233}Th and ^{40}K) in different sites in Alexandria coastline.

Elements	Locations									
	El-Max		Abu-Qir		Eastern Harbor		El-Dekhila		Rashid	
	Water	Sediment	Water	Sedimeut	Water	Sediment	Water	Sediment	Water	Sediment
H_{ex}	0.20353	0.24407	0.38659	0.28137	0.30826	0.20758	0.32797	0.26018	0.34256	0.36602
I_r	0.5942	0.7085	1.1591	0.8059	0.9273	0.5771	0.9740	0.7413	1.0296	0.9784
Absorbed dose rate ($m \cdot Sv \cdot y^{-1}$)	37.7438	33.7277	63.1650	50.1522	58.9407	37.0190	61.6391	46.7429	65.2535	67.4653
Annual absorbed dose rate ($m \cdot Sv \cdot y^{-1}$)	0.0463	0.0414	0.0775	0.0615	0.0724	0.0455	0.0757	0.0574	0.0802	0.0829

REFERENCES

1 Ibrahiem, N.M., Abd El Ghani, A.H., Shawky, E.M., Ashraf, E.M. and Farouk, M.A. (1993) Measurement of Radioactivity Levels in Soil in the Nile Delta and Middle Egypt. Health Physics, 64, 620-627.

2 Ibrahiem, N.M., Shawky, S. and Amer, H.A. (1995) Radioactivity Levels in Lake Nasser Sediments. Applied Radiation and Isotopes, 46, 297-299.

3 Hu, Q.H., Weng, J.Q. and Wang, J.S. (2010) Sources of Anthropogenic Radionuclides in the Environment: A Review. Journal of Environmental Radioactivity, 101, 426-437.

4 Hong, C. and Haux, C. (2008) Binding and Detoxification of Heavy Metals in Lower Vertebrates with Reference to Metallothionein. Comparative Biochemistry and Physiology: Part C, 100, 137.

5 Wang, J.J., Wang, C.J., Laiand, S.Y. and Lin, Y.M. (1998) Radioactivity Concentrations of ^{137}Cs and ^{40}K in Basidiomycetes Collected in Taiwan. Applied Radioactivity and Isotopes, 49, 29-34.

6 Shobier, A.H., Abdel Ghani, S.A. and Shreadah, M.A. (2011) Distribution of Total Mercury in Sediments of Four Semi-Enclosed Basins along the Mediterranean Coast of Alexandria. Egyptian Journal of Aquatic Research, 37, 1.

7 Yu, R.L., Yuan, X., Zhao, Y.H., Hu, G.R. and Tu, X.L. (2008) Heavy Metal Pollution in Intertidal Sediments from Quanzhou Bay, China. Journal of Environmental Sciences, 20, 664-669.

8 DeForest, D., Brix, K. and Adams, W. (2011) Assessing Metal Bioaccumulation in Aquatic Environments: The Inverse Relationship between Bioaccumulation Factors, Trophic Transfer Factors and Exposure Concentration. Aquatic Toxicology, 84, 236.

9 Fu, F. and Wang, Q. (2011) Removal of Heavy Metal Ions from Wastewaters: A Review. Journal of Environmental Management, 92, 407.

10 United Nations Scientific Committee on the Effects of Atomic Radiation (2000) Sources, Effects and Risks of Ionizing Radiation, Report to the General Assembly, with Annexes. United Nations, New York.

11 Oliveira Ribeiro, C.A., Schatzmann, M., Silva de Assis, H.C., Silva, P.H., Pelletier, E. and Akaishi, F.M. (2002) Evaluation of Tributyltin Subchronic Effects in Tropical Freshwater Fish (Astyanax bimaculatus, Linnaeus, 1758). Ecotoxicology and Environmental Safety, 51, 161-167.

12 Damek-Proprawa, M. and Sawicka-Kapusta, K. (2003) Damage to the Liver, Kidney and Testis with Reference to Burden of Heavy Metals in Yellow-Necked Mice from Areas around Steelworks and Zinc Smelters in Poland. Toxicology, 186, 1-10.

13 Hosono, T., Su, C., Delinom, R., Umezawa, Y., Toyota, T., Kaneko, S. and Taniguchi, M. (2011) Decline in Heavy Metal Contamination in Marine Sediments in Jakarta Bay, Indonesia Due to Increasing

Environmental Regulations. Estuarine, Coastal and Shelf Science, 92, 297-306.

14 (1986) Annual Book of American Society for Testing and Materials. Soil Sampling. Vol. 11, 1.

15 APHA (American Public Health Association) (1995) Standard Methods for the Examination of Water and Waste Water. 19th Edition, Washington DC.

16 Perin, G., Fabris, R., Manente, S., Rebello Wagener, A., Hamacher, C. and Scotto, S. (1997) A Five-Year Study on the Heavy Metal Pollution of Guanabara Bay Sediments (Rio de Janeiro, Brazil) and Evaluation of the Metal Bioavailability by Means of Geochemical Speciation. Water Research, 31, 3017-3028.

17 IAEA. International Atomic Energy Agency (2004) Sediment Distribution Coefficients and Concentration Factors for Biota in the Marine Environment. Technical Report Series, International Atomic Energy Agency, Vienna.

18 Papaefthymiou, H. and Psichoudaki, M. (2008) Natural Radioactivity Measurements in the City of Ptolemais (Northern Greece). Journal of Environmental Radioactivity, 99, 1011-1017.

19 Kcyser, R.M. (1995) Characterization and Applicability of Low-Background Germanium Detectors. Technical Note, EG&G ORTEC, Oak Ridge.

20 Hayumbu, P., Zaman, M.B., Lubaba, N.C.H., Munsanje, S.S. and Muleya, D. (1995) Natural Radioactivity in Zambian Building Materials Collected from Lusaka. Journal of Radioanalytical and Nuclear Chemistry, 199, 229-238.

21 Tarazona, J.V., Muñoz, M.J., Carbonell, G., Carballo, M., Ortiz, J.A. and Castaño, A. (1991) A Toxicological Assessment of Water Pollution and Its Relationship to Aquaculture in Algeciras Bay, Cadiz, Spain. Archives of Environmental Contamination and Toxicology, 20, 480-487.

22 Mohapatra, S.P. (1988) Distribution of Heavy Metals in Polluted Creek Sediment. Environmental Monitoring and Assessment, 10, 157-163.

23 Seshan, B.R.R., Natesan, U. and Deepthi, K. (2010) Geochemical and Statistical Approach for Evaluation of Heavy Metal Pollution in Core Sediments in Southeast Coast of India. International Journal of Environmental Science & Technology, 7, 291-306.

24 Loska, K., Cebula, J., Pelczar, J., Wiechua, D. and Kwapulinski, J. (1997) Use of Enrichment, and Contamination Factors Together with Geoaccumulation Indexes to Evaluate the Content of Cd, Cu, and Ni in the Rybnik Water Reservoir in Poland. Water, Air, and Soil Pollution, 93, 347-365.

25 Hakanson, L. (1980) An Ecological Risk Index for Aquatic Pollution Control a Sedimentological Approach. Water Research, 14, 975-1001.

26 Tomlinson, D.C., Wilson, J.G., Harris, C.R. and Jeffrey, D.W. (1980) Problems in Assessment of Heavy Metal Levels in the Estuaries and the Formation of Pollution Index. Helgoländer Meeresuntersuchungen, 33, 566-575.

27 Chakravarty, M. and Patgiri, A.D. (2009) Metal Pollution Assessment in Sediments of the Dikrong River, NE India. Journal of Human Ecology, 27, 63-67.

28 Papachristodoulou, C.A., Assimakopoulos, P.A., Patronis, N.E. and Loannides, K.G. (2003) Use of HPGe γ-Ray Spectrometry to Assess the Isotopic Composition of Uranium in Soils. Journal of Environmental Radioactivity, 64, 195-203.

29 IAEA (2004) Predisposal Management of Organic Radioactive Waste. Technical Reports Series No. 427, Vienna.

30 United Nations Scientific Committee on the Effects of Atomic Radiation (2000) Sources, Effects and Risks of Ionizing Radiation, Report to the General Assembly, with Annexes. United Nations, New York.

CHAPTER 4

Indoor Air Pollutants and the Impact on Human Health

Marios. P. Tsakas[1], Apostolos P. Siskos and Panayotis A. Siskos

[1] Laboratory of Environmental Chemistry, National and Kapodestrian University of Athens, Greece

1. INTRODUCTION

The major area of public concern and government policy, in terms of the impact of air pollution on human health, continues to be outdoor air. However, over the last two decades, indoor air quality (IAQ) has caused increasing concern due to the adverse effects that it may have on human health. The term "indoors" is used in relative literature to refer to a variety of environments, including homes, workplaces, and buildings used as offices or for recreational purposes. In addition, a number of studies have been carried out to measure various compounds inside vehicles during commuting activities. Most people in the developed world spend up to 90% of their time in an indoor environment and up to 60% of the workforce work in an office. (Tsakas & Siskos, 2010; McCurdy et al., 2000; Ashford & Caldart, 2008; Andersson & Klevard Setterwall, 1996) Decreased ventilation rates for energy conservation, along with increased use of synthetic materials in buildings, have resulted in increased health complaints from building occupants (Siskos, 2003). Many indoor pollutants are either known, or suspected to be, allergens, carcinogens, neurotoxins, immunotoxins or irritants, while all may contribute to sick building syndrome (SBS). The set of health symptoms associated with SBS includes nasal, ocular and generalised diseases. According to various studies performed in public buildings by the National Institute of Occupational Safety and Health (Soldatos et al., 2003), the three most significant symptoms that were experienced in more than 70% of the buildings are dry eyes, dry throat and headaches.

The IAQ and the presence of air pollutants in indoor environment is a worldwide issue, since many governments and environmental institutes have faced this serious phenomenon. Starting in the 1990s in Japan, tightly sealed buildings with low ventilation rates have been constructed. This, combined with the use of some new types of building materials has often resulted in IAQ problems. Many inhabitants suffering with SBS and multiple chemical sensitivity (MCS), have been reported (McCurdy et al., 2000; Zhang & Niu, 2004). As a result the Japanese Ministry of Health, Labour and Welfare have introduced indoor air guidelines for a range of VOCs including HCHO based on hazard assessments (Shinohara et al., 2009).

The importance of IAQ has also been recognised in Europe and has been identified as an important element within the European Collaborative Action (ECA) (ECA, 1998) and the European Environment and Health Action Plan (Dimitroulopoulou et al., 2006). In America, the State of California has adopted an active programme for the last two decades aiming to the reduction of indoor air pollution, which has led to a range of policy instruments (Waldman & Jenkins, 2004). Over recent years, important steps have been made towards setting IAQ standards and guidelines in the UK (Dimitroulopoulou et al., 2006) and recently, the UK Department of Health Committee on the Medical Effects of Air

Pollutants (Short, 2001), launched a guidance document on the effects of indoor air pollutants. In many developing countries, exposure to indoor air pollution causes a major health burden (Committee on the Medical Effects of Air Pollutants, 2004). Increased concern regarding indoor air quality especially in the last two decades has led to a number of studies and meetings on the subject. For example, in Greece many researchers have conducted significant studies about IAQ issue (Siskos & co-workers, 2001, 2003, 2005, 2010; Helmis & co-workers, 2007, 2009; Santamouris et al., 2001). With increasing concern in relation to health effects, in recent years the problem has come into sharper focus. Additional new sources of contaminants are being introduced, which haven't been measured before. In several countries, studies have been undertaken, in some cases involving comprehensive investigations of the factors governing air quality, so that effective control measures ranging from the setting of minimum ventilation standards, to controlling, or even banning, certain products such as urea-formaldehyde foam insulation or unvented paraffin or gas heaters. It is nevertheless recognized that some of the responsibility for maintaining acceptable and healthy indoor air quality will continue to rest with building owners and occupants of buildings.

Some studies have revealed a variety of contaminants of indoor air including odorous, non-odorous gases and vapours, and particles, and although there were suggestions that some of these contaminants could be responsible for health effects, proving causal relationships is exceedingly difficult even where elevated levels of potentially toxic substances exist (World Health Organization [WHO], 1989; Perry & Kirk, 1986; WHO, 1986; Priorities for Indoor Air Research and Action, 1991).

2. INDOOR AIR POLLUTANTS AND THEIR SOURCES

There are many indoor air contaminants, which can be separated based on their effects on human health, the frequency of their appearance, their usual concentration levels, their sources etc. This chapter is focused, primarily, on those species common to indoor and outdoor air environments and those who are measured more often in indoor environments.

2.1. Radon

The main source of indoor radon is its immediate parent radium-226 in the ground of the site and in the building materials (Nero, 1988, 1989). Outdoor air also contributes to the radon concentration indoors, via the ventilation air. Tap-water and the domestic gas supply are usually radon sources of minor importance, with a few exceptions. In most situations it appears that elevated indoor radon levels originate from radon in the underlying rocks and soils (Castren et al., 1985). This radon may enter living spaces in dwellings by diffusion or pressure driven flow if suitable pathways between the soil and living spaces are present. It should be noted, however, that in a minority of cases elevated indoor radon levels may arise due to the use of building materials containing high levels of radium-226. Examples of such materials, used in some buildings, are by-product gypsum, alum shale and volcanic tuffs.

The United Nation Scientific Committee on the Effects of Atomic Radiations (UNSCEAR) has made a very simple model to try to estimate the relative contribution of these sources: for a "typical" house, with a radon concentration of 50 Bq/m^3 at ground floor, the contributions of soil, building materials and outdoor air are, respectively, 60%, 20% and 20%, while for the upper floors in high rise buildings, where the radon concentration-is estimated to be "typically" 20 Bq/m^3, these values become 0%, 50% and 50% (UNSCEAR, 1993).

2.1.1. Soil

For those who live close to the ground, e.g. in detached houses or on the ground floor of apartment buildings without cellars, the most important radon source is radium in the ground.

The radium concentration in soil usually lies in the range 10 Bq/kg to 50 Bq/kg, but it can reach values of hundreds Bq/kg, with an estimated average of 40 Bq/kg (UNSCEAR, 1993). Typical radon concentrations in soil gas range from 10000 Bq/m^3 into 50000 Bq/m^3. The potential for radon entry from the ground depends mainly on the activity level of radium-226 in the subsoil and its permeability with regard to air flow. Example of terrains with a high radon potential are alum shales, some granites and volcanic rocks, due to high concentrations of radium-226 and the presence of eskers (gravel, sand and rounded stone deposited from subglacial streams during the ice ages), all these being characterised by high permeability. The ground could also be contaminated with waste tailings from uranium or phosphate mining operations with enhanced activity levels (Tyson et al., 1993).

The ingress of radon from the soil is predominantly one of pressure-driven flow, with diffusion playing a minor role (de Meijer et al., 1992). The magnitude of the inflow varies with several parameters, the most important being the air pressure difference between soil air and indoor air, the tightness of the surfaces in contact with the soil on the site, and the radon exhalation rate of the underlying soil. If there is no airtight layer between the basement and the ground, the underpressure indoors causes radon to be drawn in from the ground under the building. Underpressure occurs in most houses if either the adjustment of inlet and outlet of air in forced ventilation systems or the outdoor air supply for vented combustion appliances is inappropriate. The underpressure may be considerable for all types of ventilation systems when the inlet air is restricted too much. The tightness of the structures has to do with e building regulations and techniques and is very dependent on cracks, openings and joints. Structures are hardly ever so airtight that radon inflow is completely prevented. For example, to get a radon daughter concentration of less than 100 Bq/m^3 EER in a house with a volume of 500 m^3 and a ventilation rate of 0.5 air changes per hour, not more than 1 m^3 per hour must be allowed to leak into the house if the radon gas concentration in soil air is about 50000 Bq/m^3. Such values are quite typical.

2.1.2. Building Materials

Building materials are generally the second main source of radon indoors, while in the Seventies they were considered the principal one (UNSCEAR, 1977; Meyer et al., 1986). Radon exhalation from building materials depends not only on the radium concentration, but also on factors such as the fraction of radon produced through material release, the porosity of the material and the surface preparation and finish of the walls. In general, no action needs to be taken concerning traditional building materials. Typical values for radium and thorium content in building materials are 50 Bq/kg or less (Nuclear Energy Agency Organisation for Economic Co-operation and Development - NENOECD, 1979). Building materials containing by-product gypsum (UNSCEAR, 1982) and concrete containing alum shale (Swedjemark & Mjones, 1984) may have much higher radium concentrations. The activity concentrations in brick and concrete may also be high if the raw materials have been taken from locations with high levels of natural radioactivity. Examples of such natural materials, used in some buildings, are volcanic tuffs and pozzolana (Sciocchetti et al., 1983; Campos Venuti et al., 1984; Battaglia et al., 1990), where radium and thorium content can reach some hundreds of Bqlkg. Other measurements of radioactivity content and exhalation of building materials are reported inNENOECD (1979).

Building materials are the main sources of radon-220 (also called "thoron") in indoor air. Due to its short half life (55 s), thoron originating in soil in effect is usually prevented from entering buildings and therefore makes negligible contribution to indoor thoron levels. For this reason and due to the greater difficulties of measurement, thoron concentration measurements are very much fewer than those for radon. Although the indoor thoron concentrations are usually low (Cliff, 1992; UNSCEAR, 1993), in some cases the doses due to this isotope and its daughters are significant and comparable to those due to radon- 222 (Sciocchetti et al., 1983, 1992; Guo et al., 1992; Bochicchio et al., 1993; Doi & Kobayashi, 1994).

2.1.3. Outdoor Air

Outdoor air usually acts as a diluting factor, due to its normally low radon concentration, but in some cases, as in high rise apartments built with materials having very low radium content, it can act as a real source. The radon concentration in outdoor air is mainly related to atmospheric pressure, and (in case of non-perturbative weather) it shows a typical oscillating time pattern, with higher values during the night.

Until a few years ago the average level of radon gas concentrations in the atmosphere at ground level was, in most cases, assumed to be of the order of few Bq/m^3 -e.g. in the range of 4 to 15 Bq/m^3 in USA (Gesell, 1983), but more recent measurements seem to indicate higher values, reaching some tens of Bq/m^3 (Hopper et al., 1991; Robé et al., 1992; Bochicchio et al., 1993; Deyuan, 1993; Grasty 1994;Price et al., 1994). Quite high radon concentrations in the outdoor air have been reported near substantial radon sources, such as mine tailings (Tyson et al., 1993), or in the case of particular weather conditions, such as thermal inversion or very low precipitation (Grasty, 1994).

Ambient air over oceans has very low values (~ 0.1 Bq/m^3) of radon concentrations, due to the minimum presence of radium in the sea water and the high solubility of radon in water at low temperatures. Therefore radon concentration in outdoor air of islands and coastal regions is generally lower than in continental countries, e.g. United Kingdom and Japan have an average outdoor air value of ~4 Bq/m^3.

Taking into account recent measurements, the mean value of outdoor radon concentrations adopted by UNSCEAR in its last report has been changed from 5 to 10 Bq/m^3 for continental areas and somewhat less in coastal regions (UNSCEAR, 1993).

2.1.4. Tap Water

In wells drilled in rock the radon concentrations of water may be high. When such water is used in the household, radon can be partially released into the indoor air, causing an increase in the average radon concentrations. In a few regions, such as Finland and Maine (USA), the tap water from wells drilled in rock has been shown to contribute significantly to radon concentrations indoors. Radon concentrations in tap-water from deep wells can range from 100 kBq/m^3 to 100 MBq/m^3 (UNSCEAR, 1988). The indoor radon concentrations in these regions may already be high due to high rates of radon entry from the ground. The world average radon concentration in all types of water supplies is assumed to be 10 kBq/m^3 (UNSCEAR, 1993).

2.1.5. Domestic Gas

In some regions, natural gas used for cooking and heating contains elevated concentrations of radon, which is released on combustion. Normally this source is insignificant, and can be monitored at transmission and distribution points. Typically the radon level in natural gas is about 1000 Bq/m^3. Natural gas, as it is usually supplied, contains gas from a number of wells and fields and thus can vary over time, depending on the proportions supplied by different sources (UNSCEAR, 1993).

2.2. Oxides of Nitrogen

2.2.1. Nox

A large number of studies of NO and NO_2 have been carried out in many different indoor air environments (Finlayson-Pitts, 1999; Pitts et al., 1985). Because of air exchange, indoor levels are generally higher when outdoor levels increase (Hoek et al., 1989; Rowe et al., 1991; Hisham & Grosjean, 1991; Spengler et al., 1994; Weschler et al., 1994; Baek et al., 1997). However, enhanced indoor levels can be found when combustion sources are present. These include gas stoves, paraffin heaters, water heaters, and cigarette smoke (Wade et al., 1975; Marbury et al., 1988; Ryan et al., 1988;Petreas et al., 1988; Hoek et al., 1989; Pitts et al., 1989; Spengler et al., 1994; Levy et al., 1998). While combustion generates primarily NO, the focus indoors has been on NO_2 because of its health impact. Again, the use of gas stoves was highly correlated with indoor NO_2, with an indoor/outdoor concentration ratio of 1.19 for homes with a gas range compared to 0.69 for those without a gas stove. The ratio was even higher for homes with a paraffin space heater, 2.3 compared to 0.85 without such a heater (Levy et al., 1998). Both the indoor and outdoor concentrations of NO_2 were higher in cities where at least 75% of the homes had gas stoves; for example, the mean outdoor NO_2 concentration in such gas-intensive cities was 38 ± 20 ppb, compared to 14 ± 6 ppb in cities where fewer than 25% of the households had gas stoves installed. High concentrations of NO_2 have also been measured in indoor skating rinks where the use of ice resurfacing machines powered by propane, gasoline, or diesel fuel results in significant emissions (e.g., Brauer & Spengler, 1994; Brauer et al., 1997; Pennanen et al., 1997). Mean concentrations of NO_2 of ~200 ppb have been reported, with some rinks having concentrations up to 3 ppm! The indoor-to-outdoor ratios of the arithmetic mean concentrations varied from about 1 to 41, with an overall mean of 20. In the absence of such sources of NOx, indoor and outdoor concentrations are quite similar (Weschler et al., 1994), since removal of NO and NO_2 indoors, e.g., on surfaces, is relatively slow. However, as it has been discussed shortly, although the surface reaction of NO_2 is relatively slow, it is still of interest since it generates nitrous acid (HONO). Different surfaces found inside homes have been found to have different removal rates for NO_2. In short, there is a variety of evidence that there are higher levels of NO_2 indoors when combustion sources are present and that the concentrations generated indoors can be quite substantial in some circumstances. One word of caution is in order, however, particularly in regards to earlier measurements of NO_2.

2.2.2. Hono and Hno_3

HONO is formed by the reaction of NO_2 with water on surfaces. The reaction is usually represented as

$$2NO2+H2O \rightarrow HONO + HNO3 \quad (1)$$

Although the detailed mechanism is not known; gaseous HNO_3 is not generated in equivalent amounts, something which has been attributed to its remaining being adsorbed on the surface. This overall reaction occurs on a variety of surfaces in the laboratory and hence might be expected to also occur on surfaces in other environments, such as homes. This, indeed, is the case. (Pitts et al., 1985) first used differential optical absorption spectrometry (DOAS) to establish unequivocally that NO_2 injected into a mobile home forms HONO. Interestingly, the dependence of the rate of HONO generation on the NO_2 concentration was similar to that measured in laboratory systems, consistent with production in, or on, a thin film of water adsorbed on surfaces. A number of studies have confirmed that the behaviour is similar to that in laboratory systems; i.e., the rate of production of HONO increases with NO_2 and with relative humidity. Indoor levels of HONO as high as 8 ppb as a 24-h average and 40 ppb as a 6-h average have been reported in normal, in-use buildings and homes (Febo & Perrino, 1991; Spengler et al., 1993; Weschler et al., 1994). The ratio of HONO to NO_2 indoors can be quite large, up to ~0.15 (e.g., Febo & Perrino, 1991; Brauer et al., 1990, 1993; Spengler et al., 1993). This can be compared to typical values of a few percent outdoors. High levels of HONO (up to ~ 30 ppb) have also been measured in automobiles in use in polluted urban areas, and again, the ratio of HONO to NO2 was quite large, ~0.4, compared to 0.02-0.03 measured outdoors in the same study (Febo & Perrino, 1995). The generation of NO was attributed by Spicer and co-workers to a reaction of gaseous NO_2 with adsorbed HONO:

$$NO_2\,(g) + HONO(ad) \rightarrow H^{+} + NO^{-}{}_3 + NO_3\,(g) \qquad (2)$$

The same process was hypothesised to explain some time periods in a commercial office building when indoor NO actually exceeded outdoor NO (Weschler et al., 1994). As is the case in laboratory systems, equivalent amounts of HNO_3 are not observed as might be expected from the stoichiometry of reaction (1), likely due to HNO_3 remaining on the surface after formation and/or being taken up by surfaces. The accumulation of nitrate on indoor surfaces in a commercial building has been reported byWeschler and Shields (1996) and attributed to the formation and uptake of HNO_3 via reactions of $NO_3$and/or oxidation of nitrite (i.e., adsorbed HONO) in an aqueous surface film. Subsequently, it was shown that HONO is also directly emitted by gas stoves (Pitts et al., 1989). In a house used for investigating indoor air pollution that had natural gas fueled appliances (a convective heater, a radiant heater, and a range with four burners), both the surface reaction of NO_2 and the direct combustion emissions contributed significantly to the measured indoor HONO. When an appliance was operational, the contribution of direct emissions was the more important source (Spicer et al., 1993). In short, the "dark reaction" of NO_2 with water on surfaces is ubiquitous and occurs not only in laboratory systems but also indoors. The combination of this heterogeneous reaction with combustion sources of HONO can produce significant concentrations of HONO indoors. As a result, there is a concern regarding the health impacts of nitrous acid, not only because it is an inhalable nitrite but also because it is likely the airborne acid present in the highest concentrations indoors.

2.2.3. Co and So_2

As for NOx, combustion sources such as gas stoves and paraffin heaters can be significant sources of indoor CO. The ratio of indoor to outdoor concentrations of CO in homes using gas stoves has been measured to be 1.2-3.8 (Wade *et al.*, 1975), with the highest ratios found close to the source. Similarly, higher CO levels indoors compared to outdoors have been reported for restaurants in Korea, with those using charcoal burners as well as gas giving much higher concentrations (Baek *et al.,* 1997). In buildings where motor vehicle exhaust can be entrained from outdoors or attached parking garages, elevated indoor CO levels may also result (Hodgson *et al.,* 1991). On the other hand, in homes and offices where there was no direct indoor source of CO, the indoor to-outdoor ratio was about one, and sometimes less. For example, in Riyadh, Saudi Arabia, CO concentrations were measured indoors and outdoors; the indoor to- outdoor ratio varied from 0 to 2, but was typically below one (Rowe *et al.,*1989). There have been a number of measurements of CO in the "indoor environment" of automobiles. Given that cars are major CO sources in urban areas, one might expect higher concentrations of CO during commutes and this is indeed the case. Typical CO concentrations of ~9-56 ppm have been measured inside automobiles during commutes in major urban areas (Flachsbart *et al.,* 1987; Koushki*et al.,* 1992; Ott *et al.,* 1994, 1995; Dor *et al.,* 1995; Fernandez-Bremauntz & Ashmore, 1995). This can be compared to peak outdoor levels of ~ 10 ppm in highly polluted urban areas. Thus, a significant enhancement of CO inside automobiles during commutes is common. For example, Chan *et al.* (1991) report a ratio of the in-vehicle CO concentration to that outdoors of ~ 4.5 in Raleigh, North Carolina. As is the case for CO, SO_2 levels indoors and outdoors tend to be similar if there are no combustion sources indoors.

2.2.4. Volatile Organic Compounds (Vocs)

Volatile organic compounds (VOC) are ubiquitous components not only of ambient air but also of indoor air environments, including offices, commercial and retail buildings, and homes (Shah & Singh, 1988; Finlayson-Pitts, 1999). There are three sources/categories for VOC: (1) entrainment of air from outside the building, (2) emissions from building materials, and (3) human activities inside buildings. As might be expected, given the nature of the sources, a very large variety of organic compounds have been identified and measured indoors (e.g., Brown *et al.*, 1994; Crump, 1995; Kostiainen, 1995). These numbers in the hundreds of different compounds, with the particular species and their concentrations depending on the particular sources present as well as the air exchange rates. Some of the compounds associated with the three sources: entrainment from outdoors, emissions from building materials, and anthropogenic activities - are now briefly reviewed.

Entrainment of air from outdoor sources: Entrainment of outdoor air through ventilation systems brings with it the species found in ambient air. Some of them, such as HNO_3, can be removed on surfaces such as those in air conditioning systems, and hence the indoor concentrations tend to be lower than those outdoors. Others such as NO tend to have similar concentrations indoors and outdoors if there are no significant combustion sources indoors (e.g., Weschler *et al.*, 1994). In the case of hydrocarbons, the concentrations of compounds that do not have significant indoor sources tend to be about the same as the outdoor concentrations. For example, Lewis and Zweidinger (1992) measured VOC in 10 homes in winter and showed that the concentrations of ethene, benzene, 2-methylpentane, methylcyclopentane, 2,2,4-trimethylpentane, and 2,3-dimethylbutane indoors were within experimental error of those outdoors. There are, however, some specific outdoor sources that can lead to higher concentrations of certain VOCs indoors than in the general outdoor air environment. For example, gases generated in landfills or from petroleum contamination can migrate through the soil and groundwater to adjacent buildings and homes to give larger indoor concentrations, particularly in basements and crawl spaces, than otherwise expected (Moseley & Meyer, 1992; Hodgson *et al.*, 1992; Fischer *et al.*, 1996). In one such case, the total hydrocarbon concentration was measured to be 120 ppm in a crawl space beneath the floor of a school where petroleum contamination was present from adjacent sources, compared to < 80 ppb outdoors (Moseley & Meyer, 1992). Although concentrations in various rooms were lower, they were still elevated compared to outdoors, ranging from 0.13 to 3.4 ppm. The use of pesticides *outside*buildings can also lead to enhanced concentrations of these compounds indoors. For example, Anderson and Hites (1988) measured the concentrations of chlorinated pesticides indoors and found elevated levels inside, e.g., a factor of 7 times higher for y-chlordane compared to outdoor levels. One home that had the highest indoor concentrations had been treated with chlordane about a decade earlier, presumably by subsurface injection from which the pesticide migrated into the house through cracks in the basement walls. Enhanced levels of chlorpyrifos were observed indoors in homes where soil surrounding the home had been treated on a regular basis. Another source of VOC is motor vehicle emissions, which can be drawn into buildings from outdoors or parking garages (e.g., Perry & Gee, 1994; Daisey et al., 1994). For example, motor vehicles were major sources (responsible for > 75%) of 12 of 39 individual compounds measured in a dozen buildings by Daisey *et al.* (1994). Of the 12 compounds, 5 were alkanes and 7 were aromatics. Similarly, Baek *et al.* (1997) report that vehicle emissions are important VOC sources indoors in Korea during the summer in homes and offices, as has been reported in the United States (e.g., Hodgson *et al.,* 1991; Daisey *et al.,* 1994).

Building materials: Emissions associated with building materials are major contributors to indoor levels of VOC. New buildings often have higher concentrations of certain compounds compared to older buildings. For example, enhanced levels of n-dodecane, n-decanc, and n-undecane, the xylenes, and 2-propanol have been measured in new buildings, and the total VOC concentration is generally larger (by factors of 4-23) compared to established buildings (Brown *et al.,* 1994). Kostiainen (1995) identified more than 200 individual VOCs indoors in 26 houses. In addition, they compared the VOC concentrations in normal houses to those where complaints of odours or illness had been registered. A number of different VOCs were present at increased concentrations in the houses with complaints compared to the normal houses; these included a variety of aromatic hydrocarbons, methylcyclohexane, n-propylcyclohexane, terpenes, and chlorinated compounds such as 1,1,1-trichloroethane and tetrachlorethene. Carpets are a major source of VOCs in homes. For example,Sollinger *et al.* (1993, 1994) have identified 99 different VOCs emitted from a group of 10 carpet samples, and Schaeffer *et al.* (1996) identified more than 100 different VOCs emitted from the carpet cushion alone. Emissions come not only from the carpet fibres but also from the backing materials and the adhesives used to bind the carpet to the backing. As a result, the individual compounds emitted by carpets can vary substantially, depending on the carpet construction. Many of the compounds emitted are known to be used in the manufacturing processes (e.g., e-caprolactam is used in Nylon-6 production) and /or are common solvents Emissions of VOC from carpets tend to decrease with time and increase with temperature.

The dependence of VOC emissions from building materials on relative humidity is more complex, with some emissions increasing with relative humidity, but others not. For example, Sollinger *et al.*(1994) report that the VOC emissions from carpets did not change with relative humidity over the range from 0 to 45% RH. On the other hand, the emissions of formic and acetic acids from latex paints have been reported to increase dramatically with relative humidity; for example, for one paint sample the emission rate for acetic acid almost tripled when the relative humidity was changed from 4-5% to 5-23% (Reiss *et al.*, 1995b). A number of different aldehydes have been measured indoors (Crump & Gardiner, 1989; Lewis & Zweidinger, 1992; Zhang *et al.*, 1994; Daisey *et al.*, 1994; and Reiss *et al.*,1995a), some of which are directly emitted and some of which are formed by chemical reactions indoors of VOCs such as styrene. Of these, there is an enormous amount of evidence for direct emissions of HCHO from building materials. Interest in formaldehyde emissions and levels in homes and other buildings stems from its well-known health effects, which include possible human carcinogenicity and eye, skin, and respiratory tract irritation (Feinman, 1988). Formaldehyde is emitted from urea-formaldehyde foam insulation as well as from resins used in reconstituted wood products such as particleboard and plywood (Meyer and Reinhardt, 1986); urea-formaldehyde resins comprise about 6-8% of the weight of particleboard and 8-10% of mediumdensity fiberboard (Meyer and Hermanns, 1986). Other sources include permanent press fabrics (such as draperies and clothing), floor finishing materials, furniture, wallpaper, latex paint, varnishes, some cosmetics such as fingernail hardener and nail polish, and paper products (Kelly, 1996; Howard *et al.,* 1998a, 1998b).

Many measurements of HCHO have been made in indoor air environments. In conventional homes, average concentrations are typically about 10-50 ppb (Stock, 1987; Zhang *et al.,* 1994; Reiss *et al.,*1995a). Sexton *et al.* (1989) measured concentrations of HCHO in 470 mobile homes in California and found geometric mean concentrations of 60-90 ppb, although maximum values of over 300 ppb were recorded in some cases. In a similar study in Wisconsin, levels up to 2.8 ppm were measured (Hanrahan *et al.,* 1985). Higher levels are typically found in mobile homes because of the reconstituted wood products (e.g., particleboard and plywood) used in their construction. Interestingly, HCHO does not appear to be a significant product of natural gas combustion, as levels in dwellings with and without gas stoves turned on are not significantly different (e.g., Pitts *et al.,* 1989; Zhang *et al.,* 1994). Temperature is again an important determinant of HCHO levels.

Human activities: There are many sources of VOCs associated with human activities in buildings. For example, mixtures of C_{10} and C_1~ isoparaffinic hydrocarbons, which are characteristic of liquid process copiers and plotters, have been identified in office buildings in which these instruments were in use (Hodgson *et al.,* 1991). Emissions of a number of hydrocarbons and aldehydes and ketones have been observed during operation of dry-process copiers; these include significant emissions of ethylbenzene, o-, m-, and p-xylenes, styrene, 2-ethyl-l-hexanol, acetone, nnonanal, and benzaldehyde (Leovic *et al.,* 1996). Enhanced levels of acetaldehyde in an office building in Brazil were attributed to the oxidation of ethanol used as a cleaning agent (Brickus *et al.,* 1998), although levels outdoors were also enhanced due to the use of ethanol as a fuel. Pyrocatechol has been measured in an occupational environment where meteorological charts are mapped on paper impregnated with this compound (Ekinja *et al.,* 1995), and p-dichlorobenzene is observed when mothballs containing this compound are in use (e.g., Tichenor *et al.,* 1990; Chang and Krebs, 1992). Elevated concentrations of the n-C_{13} to n-C_{18} alkanes and branched-chain and cyclic analogs were measured in a building having a history of air quality complaints; the source was found to be volatilisation from hydraulic fluids used in the building elevators (Weschler *et al.,* 1990). Enhanced levels of chlorinated compounds have been observed indoors due to human activity as well. For example, increased levels of perchloroethylene have been observed from unvented dry-cleaning units (Moschandreas & O'Dea, 1995) and volatilisation of chlorinated organics such as chloroform from treated tap water can occur (McKone, 1987). Other sources include the use of household products. For example, chloroform emissions have been observed from washing machines when bleach containing hypochlorite was used (Shepherd *et al.,* 1996). It is interesting that emissions of organics associated with the use of washing machines are decreased when the machine is operated with clothes inside (Howard and Corsi, 1998). Of course, activities such as smoking result in enhanced levels not only of nicotine (e.g., Thompson *et al.,* 1989) but also of a variety of other gases associated with cigarette smoke (e.g., California Environmental Protection Agency, 1997; Nelson *et al.,* 1998). For example, using 3-ethenylpyridine as a marker for cigarette smoke,Heavner *et al.* (1992) estimated that 0.2-39% of the benzene and 2-49% of the styrene measured in the homes of smokers was from cigarette smoke. Humans emit a variety of VOCs such as pentane and isoprene (e.g., Gelmont *et al.,* 1981; Mendis *et al.,* 1994; Phillips *et al.,* 1994; Jones *et al.,* 1995; Foster*et al.,* 1996). In addition, emissions from personal care products have been observed. Decamethylcyclopentasiloxane (D5), a cyclic dimethylsiloxane with five Si-O units in the ring, and the smaller D4 analog, octamethylcyclotetrasiloxane, are used in such products as underarm deodorant and antiperspirants at concentrations up to 40-60% by weight (Shields and Weschler, 1992; Shields *et al.,* 1996). Increased concentrations of D5 have been measured in offices and are correlated to human activity, as expected if personal care products were the major source (Shields and Weschler, 1994). In some cases, increased concentrations attributable to emissions from silicone-based caulking materials were also observed (Shields *et al.,* 1996). The use of pesticides indoors can lead to very large concentrations not only of the pesticide but of the additional VOCs used as a matrix for the pesticide, which represent most (>95%) of the mass of the material as purchased. For example, Bukowski and Meyer (1995) predict that VOC concentrations immediately after the application of a fogger could reach levels of more than 300 mg m^{-3}!

2.2.5. Ozone

Because O_3 decomposes on surfaces, indoor levels are usually lower than those outdoors due to the decomposition that occurs as the air passes through air conditioning systems and impacts building surfaces (Reiss *et al.,* 1994; Finlayson-Pitts, 1999). The measured ratio of indoor-to-outdoor concentrations of ozone vary from 0.1 to 1, but are typically around 0.3-0.5 (e.g., Druzik *et al.,* 1990;Hisham & Grosjean, 1991; Liu *et al.,* 1993; Weschler *et al.,* 1989, 1994; Gold *et al.,* 1996; Jakobi & Fabian, 1997; Avol *et al.,* 1998; Drakou *et al.,* 1998; Romieu *et al.,* 1998). Buildings with low air exchange with outside air tend to have lower ratios, ~0.1-0.3 (Druzik *et al.,* 1990; Weschler *et al.,*1994; Romieu *et al.,* 1998). For example, Gold *et al.* (1996) estimate that at outdoor ozone concentrations of 170 ppb in Mexico City, the indoor-to-outdoor ratio of O_3 at a school was 0.71 ± 0.03 with thc windows and doors open, which maximised the exchange with outside air, 0.18 ± 0.02 with the windows and doors closed and the air cleaner off, and 0.15 ± 0.02 with the windows and doors closed and the air cleaner on. There are some additional sources of O_3 indoors. These include dry-process photocopying machines, laser printers, and electrostatic precipitators (e.g., Leovic *et al.,* 1996;Wolkoff, 1999). Indeed, it is not unusual to detect O_3 by its odour during operation of some copy machines and laser printers. In the "indoor environment" in cars, ozone levels tend to be significantly less than in the surrounding area. For example, Chan *et al.* (1991) report that in-vehicle O_3concentrations during commutes in Raleigh, North Carolina, were only about 20% of those measured in the local area at a fixed station. There are several contributing factors to these low concentrations. One is that NO concentrations are higher near roadways, so that O_3 is titrated to NO_2 by its rapid reaction with NO. A second is that O_3 can decompose on the surfaces of the automobile air conditioning system. A similar titration effect has been observed inside homes where there are combustion sources of NO.

2.2.6. Particles

With the epidemiological studies suggesting increased mortality associated with particles, there has been increasing interest in indoor particle concentrations compared to outdoor levels (Finlayson-Pitts, 1999). A number of studies have examined this over the years and are summarised in a review byWallace (1996). In general, if there are no indoor sources of particles, the levels indoors tend to reflect those outdoors. For example, application of a mass balance model to measurements of indoor and outdoor particle concentrations in Riverside, California, indicated that 75% of $PM_{2.5}$ and 65% of PM_{10}in a typical home were from outdoors (Wallace, 1996). Similar conclusions were reached by Koutrakis*et al.* (1991, 1992) for homes in two counties in New York. For example, they report that 60% of the mass of particles in homes is due to outdoor sources. However, the contribution to various individual elements in the particles varies from 22% for copper to 100% for cadmium. There are some differences in indoor levels of particulate matter in areas with low outdoor compared to high outdoor levels. In the case of high outdoor levels, the indoor concentrations tend to be somewhat lower than those outdoors; for example, Colome *et al.* (1992) report that the ratio of indoor-to-outdoor median concentrations of PM_{10} is 0.7 in residences in southern California. On the other hand, when outdoor levels are low, indoor levels tend to be higher Night time mass concentrations indoors tend to be smaller than those during the day, probably because of the decreased activity. Interestingly, when individuals wear personal exposure monitors to measure their actual exposure to particles, the measured mass concentrations tend to be higher than those measured with fixed monitors located indoors. A major source of increased particles indoors is cigarette smoking. (e.g. Spengler *et al.,* 1981; Quackenboss *et al.,* 1989; Neas *et al.,*1994). In addition to the contribution to the *mass* concentrations of indoor particles, cigarette smoke is of concern because of the mutagens, carcinogens, and toxic air contaminants that are emitted (Löfroth*et al.,* 1991; Chuang *et al.,* 1991; California Environmental Protection Agency, 1997; Nelson *et al.,*1998). Thus, a variety of both gaseous and particulate polycyclic aromatic hydrocarbons (PAH) and compounds (PAC) have been identified in buildings with cigarette smoke (Offermann *et al.,* 1991;Mitra & Ray, 1995). Indeed, in the homes of smokers, almost 90% of the total PAH was from tobacco smoke (Mitra and Ray, 1995). Higher levels of mutagenic particles have also been shown to be associated with indoor air containing cigarette smoke (e.g., Lewtas *et al.,* 1987; Löfroth *et al.,* 1988,1991; Georgiou *et al.,* 1991). Other significant sources identified in a number of studies are cooking, the use of paraffin heaters, wood burning, and humidifiers. For example, a study carried out under the auspices of the U.S. Environmental Protection Agency, the TEAM study (Total Exposure Assessment Methodology), indicated that an increase in PM m of ~10-20 / μg m $^{-3}$ could be attributed to cooking (Wallace, 1996). This source will obviously depend on the amount of cooking, the types of cooking, and the ventilation. For example, Löfroth *et al.* (1991) measured emissions of particles ranging from 0.07 to 3.5 mg per gram of food cooked, depending on the particular food. Baek *et al.* (1997) measured indoor and outdoor concentrations of particles in homes, offices, and restaurants in Korea and report ratios of 1.3, 1.3, and 2.4, respectively. The higher value in restaurants, even those using only gas and not charcoal, suggests a significant contribution from cooking. Paraffin heaters can be significant sources of particles under some circumstances. For example, paraffin heaters were reported to contribute to indoor $PM_{2.5}$ in homes in Suffolk County, New York, but not Onondaga County; wood stoves and fireplaces and gas stoves did not contribute in either case (Koutrakis *et al.,* 1992; Wallace, 1996). A similar conclusion was reached in a study of eight mobile homes in North Carolina (Mumford *et al.,*1990). However, it should be noted that even where paraffin heaters do not contribute significantly to particle *mass* concentrations, they may still be important in terms of health effects. This is because of the composition of the particles emitted, which include polycyclic aromatic compounds and other mutagenic species, as well as sulfate (Traynor *et al.,* 1990). For example, Traynor *et al.* (1990) studied the emissions from unvented paraffin space heaters and identified a number of PAHs (naphthalene, phenanthrene, fluoranthene, anthracene, chrysene, and indeno[c,d]pyrene) and nitro-PAHs (1-nitronaphthalene, 9-nitroanthracene, 3-nitrofluoranthene, and 1-nitropyrene), in addition to a host of other gaseous species.

Baek *et al.* (1997) also reported increased levels of a number of gases indoors in homes and offices in Korea due to the use of paraffin heaters. In studies of indoor air in eight mobile homes, Mumford *et al.*(1991) identified the PAHs and nitro- PAHs measured in emissions from paraffin heaters by Traynor *et al.* (1990), as well as a number of compounds that may be animal carcinogens, such as cyclopenta[c,d]pyrene, benz[a]anthracene, benzofluoranthenes, benzo-[a]pyrene, and*benzo[ghi]perylene.* While the mass concentrations of PM10 did not increase with the paraffin heater on in six of the eight homes studied, the particles in five of the homes had increased mutagenicity using TA98 with or without $9 added. In short, not only the mass emissions but also the nature of the compounds emitted must be taken into account in assessing the health effects of indoor particles.

Where indoor heating and cooking involves the use of coal or biomass, indoor particle concentrations can be extremely large. For example, Florig (1997) and Ando *et al.* (1996) report that in China typical indoor total suspended particle (TSP)concentrations can be in the range from 250 to 900/μg m^{-3} in homes using coal and 950-3500 / μg m^{-3} in those using biomass fuels. These levels can be compared to annual average outdoor concentrations of 250-410 / μg m^{-3}. The high concentrations associated with coal burning combined with the mutagenic nature of the emissions have been suggested to be responsible for enhanced lung cancer in China (Mumford *et al.,* 1987). Similarly, Davidson *et al.* (1986) measured TSP concentrations of 2900-42,000/ μg m^{-3} in homes in Nepal that used biomass fuels, compared to outdoor levels of 280 / μg m^{-3}. For particles with diameters less than 4 /μm, the levels ranged from 870 to 14,000/ μg m^{-3}. Similar conclusions regarding the relative indoor and outdoor concentrations have been reached in studies of office and commercial buildings. For example, Ligockiet *al.* (1993) measured indoor and outdoor concentrations of particles and their components at five museums in southern California. The indoor-to-outdoor ratios of particle mass varied over a wide range, depending to a large extent on the ventilation and filtration systems in use. Ratios varied from 0.16 to 0.96 for particles with diameters less than 2.1/μm and from 0.06 to 0.3 for coarse particles with diameters greater than this.

2.2.7. Microbial Pollutants

Microbial pollution is a risk to health and is associated with allergic illnesses. Published results indicate that 20% of the population can be sensitised by airborne fungal spores in the UK, while 40% of the inspected houses in Germany suffer from mould-related problems (Waubke & Kusterle, 1990). The medical consequences of immune response, allergic reactions, endotoxins, mycotoxins, and epidemiology have been extensively studied by Miller (1990), Morey (1990), Gravensen et al. (1990) and Burge et al. (1990). Similarly, Legionnaires ' disease and Pontiac fever are associated with wet cooling towers and domestic hot-water systems in complex buildings.

Accordingly to the official published figures, some 560,000 people need treatment because of indoor pollution due to mites and mould in damp houses (House of Commons Environment Committee, 1991). Indoor airborne allergic components come from two sources: outdoor air-borne spores moving inside and allergic components originating inside the dwelling. The source of biological growth within buildings is associated with moisture and the formation of microclimates; it also depends upon the type of the buildings and their ventilation. Mould fungi thrive on surfaces on which there is nourishment and suitable humidity, for example on damp water pipes, windows and walls in kitchens and bathrooms, in central air-conditioning systems, circulation pumps, blowers, ventilation ductwork and air filters, central dehumidifiers, and inside damp structures. Allergenic substances can be airborne and inhaled, such as pollen, fungus and dust, digested, such as mouldy food or drink. Investigations suggest that airborne allergies cause more problems throughout the world than all other allergies combined. Additionally, cross-infection from patient to patient is of great concern in hospitals. The medical field that treats allergies recognises the following allergenic diseases: asthma, allergic rhinitis, serous otitis media, bronchopulmonary aspergillosis, and hypersensitivity pneumonitis.

Allergic load and cocktail effect: For some people, an allergic reaction in the indoor environment may be triggered by non-biological factors, such as chemicals or other indoor air pollutants, emotional stress, fatigue or changes in the weather. These factors burden allergic people further if they are suffering from allergic reactions to biological contaminants. This combination is known as 'allergic load'. Microbial contaminants propagated within the health care establishment are particularly aggressive to patients due to reduced immune system resistance.

Recently, attention has been focused on the cocktail effect of chemicals present in indoor air. Volatile organic compounds may be produced from the use of wood preservatives and remedial timber treatment chemicals, moth-proof carpets, fungicides, mouldicide-treated paints, furnishing materials such as particle board and foamed insulation which may emit formaldehyde. Biological pollutants alone or in synergetic effect with any of the above-mentioned volatile organic compounds may produce symptoms such as stuffy nose, dry throat, chest tightness, lethargy, loss of concentration, blocked, runny or itchy nose, dry skin, watering or itchy eyes or headache in sensitive people. The 'sick building syndrome' (SBS) or tight building syndromes may arise from a variety of causes. Because of the uncertainties about the causes of SBS and the rising levels of health related problems in buildings there is an increasing use of the term building-related illness (BRI) to cover a range of ailments which commonly affect building occupants.

2.2.8. Asbestos and Manmade Mineral Fibres

Asbestos is known to cause a number of diseases after occupational exposure (Brown & Hoskins, 1993). Before the hazards associated with the inhalation of these mineral fibres were understood these exposures were often very large with frequent reports of dust clouds so great that visibility in the workplaces was considerably reduced. This type of exposure is quantitatively quite different from those in the general environment that have provoked a response which in some quarters approaches hysteria. In the USA at least there is massive expenditure on asbestos removal, management and litigation.

Asbestos is a collective, trivial, name given to a group of highly fibrous minerals that are readily separated into long, thin, strong fibres occurring on sufficient large bulk deposits for their industrial exploitation. Asbestos minerals were usually used for their insulating properties, or in a composite, where they added strength, as in cement, or increased friction, as in brake shoes. Chrysotile, or white asbestos has counted for over 90% of the world trade in asbestos minerals. It is a serpentine mineral while the others (amosite (brown asbestos); crocidolite (blue asbestos); anthophyllite; tremolite; and actinolite) are all amphibole minerals. Amphibole asbestos has grater acid and water resistance than chryusotile and was used where these properties made it more suitable. Sometimes users would be unaware of the differences between the types of asbestos and so different minerals could have been used for a single application.

Recently the concern over the health effects of asbestos has been extended to another group of fibrous materials- the man-made mineral fibres (MMMF). While this term is self-explanatory a variety of types are produced with diverse chemical compositions, properties and uses. While sometimes referred to as 'asbestos substitutes' the majority of uses for the manmade fibres are relatively novel and ones for which the natural fibres are unsuitable. For example refractory ceramic fibres are resistant to considerably higher temperatures than are any of the natural fibres. The development of synthetic fibrous insulation materials has been given a great impetus in recent years by the need for more thermally efficient buildings and industrial processes.

MMMF can be made from most types of glass, from rock such as basalt, diabase and olivine and from various types of slag. Ceramic fibres can be made from kaolin or from pure silica and other oxide starting materials. The MMMF have been classified into four broad groups based on the manufacture and use: continuous filament glass fibre made by extrusion and winding processes, insulation wool (including ceramic fibre), and special purpose fibres. The non-continuous fibres are made by dropping molten material onto spinning disks or by air or steam jet impingement on a stream of the molten material. They contain a wide range of fibre sizes and are contaminated by small glassy balls called shot which often account for 50% of the product by weight.

3. FACTORS THAT INFLUENCE EXPOSURE TO INDOOR AIR POLLUTANTS

3.1. General

The types and quantities of pollutants found indoors vary temporally and spatially. Depending on the type of pollutant and its sources, sinks and mixing conditions, its concentration can vary by a factor of 10 or more, even within a small area.

Human mobility constitutes an important kind of complexity in the determination of exposure to air pollutants. Human activity patterns differ between midweek and weekend, between one season and another, and between one part of one's life and another. Activity patterns determine when and how long one is exposed to both indoor and outdoor pollutants. Therefore, in reviewing the factors that influence air-pollution exposures, we have specifically separated them into two major components: time (activity) and concentration (location).

Information on the time spent in various activities is summarised first, and then the variations in concentration often encountered in different locations. Unfortunately, most of the studies discussed were not longitudinal and thus do not offer information on seasonal differences in time spent indoors and outdoors or on regional differences in activity patterns.

Outdoor concentrations of pollutants and rates of infiltration affect the concentrations to which people are exposed indoors. Building construction techniques, as they vary geographically, and their effect on pollution infiltration are particularly important. But the measurement techniques available are limited; the need for additional studies is discussed. The rates of infiltration on a neighbourhood scale have been studied by only a few researchers. Although their work has focused on energy conservation, their findings can easily be applied to the study of impact on indoor pollution.

Patterns of human behaviour and activity determine the time spent in any specific location, and thus knowledge of them is essential in estimating exposures of populations to pollutants. As indicated byOtt (Ott, 1995), a large number of variety of studies in which data on human activities were collected from population samples have been completed over the past 50 ye.

When one examines the literature on human activities, the term "time budget" ("zeitbudget", "budget de temps") is encountered often. A time budget produces a systematic record of how time is spent by a person in some specified period, usually 24 h. It contains considerable detail on a person's activities; including the locations in which the activities take place (Michelson, 1973).

One way of obtaining time budget information from the populations surveyed is to ask each respondent to maintain a diary of his or her activities over a 24-h period or longer. In another approach, the so-called "yesterday" survey approach, the interviewer asks each responder about his or her activities on the preceding day.

Several summaries of the historical development of time-budget research have been published (Chapin, 1974; Converse, 1968; Ottensman, 1972). Ott (Ott, 1995) discussed the literature on activity patterns in the context of estimation of exposure to air pollution. Owing to the small number of field monitoring studies, the geographic distribution of indoor air pollutants has not been determined. However, it is instructive to review the geographic distribution of the major factors that affect variations in the concentrations of pollutants and their impact on the quality of the indoor environment. Outdoor air quality, air-infiltration rates, and sources of emission of indoor air pollutants are the major factors. Outdoor air quality has been studied with respect to some pollutants, and the geographic distribution of these few pollutants is well understood. Descriptive statistics published annually by EPA and state and local air-quality agencies furnish much scientific information useful in discerning regional and local differences in concentrations of carbon monoxide, total suspended particles, ozone, NOx, sulfur dioxide, sulfates, and others. It should be noted that the geographic distribution of some criteria pollutants has been studied and is easily accessible from the literature; information on non-criteria pollutants is sparse and often collected and analyzed by questionable methods.

Concentrations of chemically non-reactive pollutants in residences generally correlate with those outdoors. Distribution of indoor air quality is extremely difficult to describe on a geographic scale, because indoor air quality is determined by complex dynamic relationships that depend heavily on occupant activity and highly variable structural characteristics. Weather, which has a regional character, influences indoor air concentrations of some chemicals, such as formaldehyde, and biologic contaminants, such as bacteria and molds. Therefore, the influence of relative humidity and other weather-related conditions affecting indoor environmental quality needs to be studied geographically. Research specifically addressed to geographic distribution of indoor air quality is needed.

Typically, the air-infiltration rate for American residences is assumed to be 0.5- 1.5 ach. This assumption is supported by the results of several energy and air-quality studies that experimentally determined the range of ventilation rates for typical residences to be between 0.7 and 1.1 ach (Moschandreas & Morse, 1979). However, the sample that yielded the data is small, and statistical documentation for such statements is not strong.

The quality of indoor air is a function of outdoor air quality, emission from indoor sources, air-infiltration rates, and occupant activity is likely to vary within each metropolitan and suburban area, is indeed within each neighbourhood. Within a metropolitan area, it has been shown that an urban complex leads to the so-called urban heat reservoir (American Society of Heating, Refrigerating and Air-Conditioning engineers. ASHRAE, 1972). Urban characteristics-- such as city size, density of buildings, and population-- correlate with such meteorological factors as temperature, pressure and wind velocity (Gibson, & Cawley, 1977; Kostiainen, 1995). The urban heat island affects both urban pollution patterns and meteorological characteristics that affect the infiltration rates of buildings. Thus, although the exact nature of the impact on indoor air quality is not known, it is fair to expect that the heat island to have an impact on the indoor environment that is likely to be adverse. Also, the variations due to mechanical ventilation, structural differences, and air infiltration may vary within a neighbourhood as a function of such factors as house orientation, tree barriers, and terrain roughness.

Occupant activity, air-infiltration rates, the indoor sources of pollutants and their chemical natures are some of the factors that cause variations within a city. A study (Moschandreas et al., 1980) in the Boston metropolitan area obtained indoor air samples from 14 residences under occupied "real-life" conditions for 2 week each. The indoor air character not only was driven by outdoor concentrations, but was greatly affected by other factors, such as indoor activities.

Wind speed, temperature difference, pressure differential, terrain characteristics (roughness and barriers, such as trees and fences), building orientation, and structure characteristics may be affected by the location of one residence relative to another within a neighbourhood.

The indoor air quality of an individual building is often characterised by the 24-h average for the concentration of one pollutant measured at one sampling location. Because the activity patterns of persons are such that more time is spent in some indoor areas than in others, the question arises (Moschandreas et al., 1978): "Do indoor zones (independent areas) with distinct pollutant patterns exist?" At issue here is whether sampling from one monitoring zone is sufficient to characterise the air quality of an entire building.

In an extensive analytic study of indoor air quality, Shair and Heitner (1974) assumed that there are no pollutant gradients in the indoor environment. The experimental database of Moschandreas and co-workers (1980) verified that the gradients in concentrations of several gaseous pollutants in the residential environment are negligible. J.D. Spengler, R.E. Letz, J.B. Ferris, Jr., T. Tibbets, and C. Duffy reported (at the annual meeting of the Air Pollution Control Association, 1981) on weekly nitrogen dioxide measurements in 135 homes in Portage, Wisconsin. On the average, kitchen concentrations were twice those in bedrooms in homes that had gas stoves. A study of the air quality in a scientific laboratory by West (1977) showed an almost uniform distribution of an intern tracer continuously released in the room. Similar experiments performed by Moshandreas et al. in residential environments showed that equilibrium is reached throughout a house within an hour. Episodic release of sulphur hexafluoride tracer gas also illustrates this point. The source location was the living room; adjacent locations were the kitchen and the hall. Episodic release of this inert gas in 24 residences was followed by uniform indoor distributions within 30 min (Moschandreas et al., 1978; Peterka & Cermak, 1977). The one-zone concept does not require instantaneous mixing, because it is based on the behaviour of hourly average pollutant concentrations.

Moschandreas and associates (1980) used a different database derived from the monitoring of 14 indoor environments in the Boston metropolitan area. Analysis of variance was used to reach the following conclusions:

- Pollutants (ozone and sulphur dioxide) generated principally outdoors have little or no interzonal statistical difference indoors.
- Pollutants with strong indoor generation have interzonal statistical differences in residences with gas facilities and offices, but not in electric-cooking residences. In general, the observed differences are not large, and the health differences are not expected to be serious.
- Depending on indoor activity and outdoor episodic pollutant activity, the indoor arithmetic 24-h average may or may not adequately represent the variation of hourly indoor concentrations.
- Although more than one zone would be preferable, hourly pollutant concentrations obtained from one indoor zone adequately characterise the indoor environment.

The most important factors that influence exposure to indoor air pollutants are the one described under. It should be noticed that these conclusions are not applicable to short-lived pollutants. Contaminants associated with tobacco smoke, bathroom odours, allergens, and other pollutants related to dust are expected to vary considerably in a given residence. Additional documentation is needed to determine the extent of this variation.

3.2. Site Characteristics

The characteristics of a building site that influence indoor air quality are addressed as three related subjects: air flow around buildings, proximity to major sources of outdoor pollution, and type of utility service available.

The air flow around a building has been shown to be determined by the local characteristics of the geometry of surrounding buildings (Peterka & Cermak, 1977), the location and type of surrounding vegetation

(White, 1995), the terrain (Geiger, 1965), and the size and shape of the building itself. Pollutants can be transferred by the air flow from the street level, over the façade of the building and onto the roof (Cermak, 1976). Field tests of isolated buildings have been used to develop scaling coefficients for both isothermal and and stratified cases of surface wind pressures, turbulence, and dispersion (Davenport, 1960.). Air flow around the building creates low pressure on the leeward side and/or the sides adjacent to the windward face, as well as the roof. Air pollutants released from stacks, flues, vents, and cooling towers in the region can re-enter the building through make-up air intakes for ventilation (Cermak, 1976).

Trees and forests have been generally studied as shelter belts in an agricultural context. Shelter belts affect air flow around buildings. When an air current reaches a shelter belt, part of it is deflected upward with only a slight change in velocity, part passes through the crowns of the trees with very low velocity, and part is deflected beneath the canopy with rapidly decreasing velocity (Federer, 1971). The changes in velocity of air flow outside may change the infiltration rate and thus affect indoor air quality.

The location of a building relative to a major outdoor pollution source can affect indoor air quality. For example, buildings near major streets or highways often have high carbon monoxide and lead concentrations, owing to the infiltration of these pollutants.

The type of utility service available is also related to the site of the building and may affect the character of its indoor environment. The availability of particular fuels (e.g., natural gas and oil) influences the types and concentrations of pollutants (e.g., combustion products) emitted by space-and water- heating. Service moratoria, development timing, and development scale are institutional elements that contribute to the variability of utility services and thus can affect indoor air quality.

3.3. Occupancy

Occupancy factor that affect indoor air quality include the type and intensity of human activity, spatial characteristics of a given activity, and the operation schedule of a building.

Several human activities-such as smoking, cleaning and cooking-generate gaseous and particulate contaminants indoors. The number of occupants of a space and the degree of their physical activity (i.e., metabolic rate at rest or under intense activity) are related to the production of various pollutants, such as carbon dioxide, water vapour, and biologic agents. If the only source of indoor carbon dioxide is that caused by occupants, ventilation rates may be proportional to the number of people and their metabolic rates (Mclntyre, 1980). Although studies have shown no constant relationship between carbon dioxide concentrations and the concentrations of other pollutants, carbon dioxide concentration is often used as a general indicator of the adequacy of ventilation in an occupied space.

Building occupancy is often expressed as occupant density and the ratio of building volume to floor area. The importance of occupancy in indoor air quality is illustrated by the fact that the choice of natural or mechanical ventilation is based on occupant density and the spatial characteristics of the building under consideration.

Occupancy schedule and associated building use may affect the type, concentration, and time and space distribution of indoor pollutants. Because most buildings are unoccupied for substantial portions of each day, the manipulation of "operating schedule" is a means of controlling energy use (American Institute of Architects Research Corporation. Phase Two Report for the Development of Energy Performance Standards for New Buildings, 1979). Efforts to conserve energy through the design of ventilation systems can result to the degradation of indoor air quality. However, detailed studies relating ventilation capacity, occupancy schedules, energy requirements, and indoor air quality have only recently been implemented.

3.4. Design

Elements of building design that affect the indoor environment include interior-space design (space planning), envelope design, and selection of materials.

The evolution of space planning in many building types has resulted in flexibility in assigning functions to specific locations. However, this flexibility is accompanied by a decrease in the ability to predict exposure to air pollutants. In particular, "open-plan" offices and schools have serious technical problems of redundant service distribution, limited acoustic control, incomplete air diffusion, and incomplete pollutant dispersion indoors, compared with "fixed-plan" floor layouts.

Evaluation of the success of a floor plan in achieving space efficiency, structural economy, and energy efficiency is usually in terms of net area per occupant and ratio of net usable area to total area. Explicit planning for environmental quality must be included to ensure that spatial arrangements are acceptable to the occupants.

A building's structural envelope consists of both primary elements - foundations, floors, walls, and roofs- and secondary "skin" elements - facings, claddings, and sheathing. To various degrees, the function of these is to maintain the integrity of the structure under the stresses caused by structural load, wind pressure, thermal expansion, precipitation, earth movement, and fire. The integrity of the building envelope is a major consideration in uncontrolled air movement into and out of the building –usually referred to as "infiltration". This is a major factor in indoor air quality. There has been no systematic survey of infiltration rates of buildings in the United States. The dominant factor in determining a building's infiltration rate is the total area of effective leakage, as measured with fan pressurisation. Following the leakage area in importance are the terrain and shielding near the building, the mean climatic conditions during heating (or cooling) periods, and the building height (Sherman, 1981). There is much evidence (Dickerhoff et al., 1980), both in the United States and in

Europe, that houses in mild climates are "very leaky", whereas houses in severe climates are "tight".

Greater height of a building increases the "stack effect", or updraught, and exposes the building to higher wind speeds. Thus, higher wind pressures drive air through existing openings, referred to as "leakage", increasing the infiltration rate.

The dominant building factors that determine infiltration have not been identified, but a catalogue of leakage openings found in typical structures is as follows:

- Walls: Leakage around sill plates (the openings at the bottom of wallboard), electric outlets, plumbing penetrations, and headers in attics for both interior and exterior walls.
- Windows and doors: Window type is more important than manufacturer in determining window leakage. This source of leakage tends to be overrated; it contributes only about 20% of the total leakage of a house
- Fireplaces: This includes dampers, glass screens, and fireplace caps.
- Heating and cooling systems: The variables include combustion air for furnaces, dampers for stack air draft, air-conditioning units, and location of ductwork.
- Vapour barrier and insulation penetrations.
- Utility accesses: This includes recessed lighting and plumbing and electric penetrations leading to attic or outside.
- Terminal devices in conditioned space: This includes leakage of dampers, especially those for large air-handling systems.
- Structural types: Examples are drop ceilings above cupboards or bathtubs, prism-shaped enclosures over staircases in two-story houses, and elevator and utility shafts that lead from basement to attic.

Wall and ceiling materials and floor finishes are the constituents of the building interior. Modular components, weight, strength, thermal insulation, thermal stability, sound insulation, fire resistance, ease and speed of installation and ease of maintenance are among the criteria considered in the selection of materials for walls, ceiling and floors. But emphasis on first cost, ease of installation, maintenance and long service life has also led to the use of materials that may be sources of indoor contaminants.

3.5. Operations

Depending on the type of ownership (owner-occupied or developer-owned), building operation may vary considerably, and this variation may have an impact on indoor air quality. "Building operation" pertains to the following elements of a building: the building envelope, service and plant, building facilities, equipment and landscaping. Cleaning, preventive maintenance, and replacement and repair of defects are also included in building operation. The staff responsible for building operation includes management, engineering, and custodial personnel. The care responsibilities are operation of the heating, ventilation, and air-conditioning systems and building services, such as hot water, lighting and power distribution. Building operation has an impact on indoor air quality in numerous ways, but the magnitude of this impact is not known.

4. HEALTH EFFECTS OF INDOOR AIR POLLUTION

Indoor air pollution, apart from the health impact, has socio-economic costs. The potential economic impact of poor indoor air quality is quite high, and has been estimated to be in the order of tens of billions of ECU per year in Western Europe. This includes costs of medical care, loss of income during illness, days lost due to illness, poor working performance and lower productivity. Labour costs are significantly greater per square metre of office space than energy and other environmental control costs (ECA, 1989). In the US, the loss in productivity for each employee which is attributable to IAQ problems is currently estimated to be 3% (14 minutes/day) and 0.6 added sick days annually. Other estimates have been made by calculating the impact of IAQ on productivity. For instance, in Norway, the authorities estimate that the costs to society related to poor IAQ are in the order of 1 to 1.5 billion ECU per year or about 250 - 350 ECU per inhabitant. This estimation only includes costs related to adverse health effects requiring medical attention and does not include reduced working efficiency or job-related productivity losses. Thus, from an economic consideration, remedial action to improve indoor air quality is likely to be cost effective even if an expensive retrofit is required.

As far as it concerns the health effects on IAP, it is very interesting to present the methods of studying health effects, the criteria for the assessment of the impact of IAP on the community and the diverse effects of IAP on human health (ECA, 1991).

4.1. Methods Of Studying Health Effects

Methods of studying health effects of indoor pollutants can be grouped into three broad categories:

a) Human studies, subdivided into observational and experimental studies.

Epidemiological studies of pollutants are mostly observational, i.e. the investigator has no means of experimentally exposing humans to pollutants, or of allocating subjects to exposed and unexposed groups. Critical issues are therefore the validity and precision of exposure assessment, and the control for confounding factors in these studies. Recent developments have stressed the importance of reducing exposure misclassification, and of studying restricted, well defined, homogenous populations to address these issues. The main advantage is that humans are studied under realistic conditions of exposure. By themselves, observational epidemiological studies are not usually sufficient to support causality of an observed association, so that additional information is needed from other types of studies. Experimental studies are among these; however, these are only suitable for studying moderate, reversible, short term effects in persons who are healthy or only moderately ill. Their main advantage is that exposure conditions and subjects election are under the control of the investigator.

b. Animal studies, which can be subdivided into a number of categories depending on their length (acute, subchronic, chronic) or end-point (morbidity, mortality, carcinogenicity, irritation, etc.). Here, the investigator has full control over exposure conditions and health effects studied. However, the principle limitations lie in the fact that extrapolation from the studied animal species to man is always necessary. Also, while in human populations health effects with low incidences are often of interest (e.g., specific cancers), it is not feasible to study very large groups of animals to detect these low incidences. In practice, therefore, animal experiments are often carried out using very high experimental doses to compensate for the relatively small number of animals used and as a consequence, an additional extrapolation from high to low doses is also often necessary.

c. In vitro studies, in which effects of pollutants on cell or organ cultures are studied. These studies have the advantage that they are less costly than animal studies, and that results can generally be obtained in a shorter period of time. They are useful for studying mechanisms of action, but it is not usually possible to predict effects on whole organisms from their results in a quantitative way.

4.2. Criteria for The Assessment of The Impact of Iap on The Community

The process of risk characterisation for indoor pollutants occurs through several phases: hazard identification, exposure assessment, dose-effect evaluation, and finally qualitative and quantitative risk assessment. The final product of this process may be an individual risk estimate per exposure unit or the evaluation of the incidence of the concerned effects in a given population. The risk characterisation through a multi-stage process as described above is particularly informative because, by dividing the analysis of the scenario of each pollutant into steps, it allows the separate recognition of the importance of each variable in the scenario and the prediction of the changes of frequency or severity of effects obtainable by modifying (increasing or decreasing) exposure.

For some types of IAP, our understanding of human health risk is well defined. For most indoor air pollutants, however, the risk assessment process has its limitations.

First, it has been applied successfully only to individual pollutants for which information is available for exposure and dose-response relationships and for which the effect is clear, certain, and measurable, such as mortality and cancer. Little progress has been made in applying the risk assessment process to environmental issues involving pollutant mixtures or effects for which the causes are difficult to ascertain precisely, such as in heart disease, allergic reactions, headache, and malaise. A different approach is needed for the assessment and characterisation of the risks associated with most indoor air pollutants.

A basic and simple criterion for assessing the importance of the health risk related to indoor pollution makes reference to the severity of the effect concerned and to the size of the population affected. Important issues for the community may come from severe health impacts, particularly when affecting a large segment of the population. Minor impacts, such as those related to discomfort or annoyance may, however, become important when a large number of individuals in the community are concerned.

4.3. The Impact of Iap on Humans' Health

4.3.1. Respiratory Health Effects Associated With Exposure To Iap

Several effects on the respiratory system have been associated with exposure to IAP. These include acute and chronic changes in pulmonary function, increased incidence and prevalence of respiratory symptoms, augmentation of pre-existing respiratory symptoms, and sensitisation of the airways to allergens present in the indoor environment. Also, respiratory infections may spread in indoor environments when specific sources of infectious agents are present, or simply because the smaller indoor mixing volumes allow infectious diseases to spread more easily from one person to the next. The latter mechanism is particularly operative in schools, nursery schools, etc.

Observed changes in pulmonary function due to exposure to, e.g., tobacco smoke in the home, have mostly been due to acute or chronic airway narrowing leading to obstruction of air flow. This is measured as a reduction in the quantity of air that can be exhaled in one second after deep inspiration (FEVI), and a limitation in the various measures of air flow such as Peak Expiratory Flow (PEF), Maximum Mid Expiratory

Flow (MMEF), and Maximum Expiratory Flow at x% of Forced Vital Capacity (MEFx). In growing children, it has also been suggested that lung development could be impaired by exposure to IAP.

Asthma, manifested by attacks of excessive airway narrowing leading to shortness of breath and wheezing, can be caused or aggravated by exposure to allergens at home, but it has also been associated with exposure to substances such as nitrogen dioxide and environmental tobacco smoke (ETS). Bronchitis, manifested in inflammatory changes in the airways and mucus hypersecretion has been linked to high levels of ambient air pollution in the past, and to exposure to ETS in the home in recent studies. Respiratory symptoms which have been associated with exposure to indoor air pollutants are symptoms mostly related to the lower airways such as cough, wheeze, shortness of breath and phlegm.

In contrast to the occurrence of chemical pollutants in indoor air, attention to which has grown considerably over the past two decades, the role of infectious agents in indoor air has been known for a long time. Infectious agents can be involved in the inflammatory conditions rhinitis, sinusitis, conjunctivitis and sinusitis, in pneumonia, in asthma and in alveolitis.

4.3.2. Allergic Diseases Associated With Exposure To Iap

Allergic asthma and extrinsic allergic alveolitis (hypersensitivity pneumonitis) are the two most serious allergic diseases caused by allergens in indoor air. Allergic rhinoconjunctivitis and humidifier fever are other important diseases; it is not clear if or how the immunological system is involved in humidifier fever.

Allergic asthma is characterised by reversible narrowing of the lower airways. Pulmonary function during an attack shows an obstructive pattern in serious cases together with reduced ventilation capacity. Allergic asthma may be caused by exposure to indoor air pollutants, either acting as allergens or as irritants. Immunological specific IgE sensitisation to an airborne allergen is a major component of this disease, but non-specific hypersensitivity is also important for the asthmatic attacks occurring on exposure to irritants in the indoor air.

The prevalence of asthma varies considerably from country to country. Although asthmatic attacks seldom lead to death, the costs of medical care are considerable in terms of hospital admissions, medication, and lost work days.

Allergic rhinoconjunctivitis is also an IgE-mediated disease, but while asthma occurs in all age groups, allergic rhinoconjunctivitis is especially prevalent among children and young adults. The main symptoms are itching of the eye andlor the nose, sneezing, watery nasal secretion and some stuffiness of the nose. The severity of the symptoms varies with the exposure to the allergen. Individuals often suffer from both allergic asthma and allergic rhinoconjunctivitis and are seldom sensitive to only one allergen. Aeroallergens from house dust mites, pets, insects, moulds, and fungi in the indoor air have been shown to be associated with allergic asthma and/or rhinoconjunctivitis. Extrinsic allergic alveolitis, also called hypersensitivity pneumonitis, is characterised by recurrent bouts of pneumonitis or milder attacks of breathlessness and flu-like symptoms. Studies of the pulmonary function during an acute episode will usually show a restrictive pattern with a decreased diffusion capacity. The disease is believed to be an inflammatory reaction in the alveoli and bronchioles involving circulating antibodies and a cell-mediated immunological response to an allergen. For example it occurs in farmers as a result of handling mouldy hay ("farmer's lung") and in pigeon breeders due to bird droppings. However, the disease has also in a few cases been associated with exposure to IAP, most frequently related to humidifiers in homes and offices contaminated with bacteria, fungi, or protozoans.

Allergic asthma and extrinsic allergic alveolitis resolve with cessation of exposure to the allergen, but continued exposure in sensitised patients may result in permanent lung damage and death from pulmonary insufficiency.

Humidifier fever is a flu-like illness involving the immune system, in which X-ray abnormalities are usually absent. The exact cause is not clear. The disease may occur among persons exposed to humidification systems contaminated with microbial growth. The symptoms typically occur 4-8 h after the exposure on the first day back at work after a weekend, but resolve within 24 h. Despite continuous exposure the disease does not recur until after the next weekend. Even though pulmonary changes are seen during attacks of humidifier fever, the disease does not lead to permanent lung damage.

4.3.3. Cancer And Effects On Reproduction Associated With Exposure To Iap

Lung cancer is the major cancer which has been associated with exposure to IAP (radon or ETS). Asbestos exposure has been linked to cancer in workers and also in workers' family members, presumably due to asbestos fibres brought into the home on workers' clothing. However, there are no studies associating asbestos exposure in homes or public buildings from asbestos used as a construction material to the development of cancer. Effects on human reproduction have been associated with exposure to chemicals in the environment, but it is as yet unclear to what extent (if any) exposure to IAP is involved

4.3.4. Sensory Effects and Other Effects On The Nervous System Associated With Iap

Sensory effects are defined as the perceptual response to environmental exposures. Sensory perceptions are mediated through the sensory systems and result in a conscious experience of smell, touch, itching, etc. Sensory effects are typically observed in buildings with indoor climate, problems because many chemical compounds found in the indoor air have odorous or mucosal irritation properties. Most indoor air chemicals with a measurable vapour pressure will be odorous when the concentration is high enough.

Sensory effects are important parameters in indoor air quality control for several reasons. They may appear as: (1) adverse health effects on sensory systems (e.g., environmentally-induced sensory dysfunctions); (2) adverse environmental perceptions which may be adverse per se or constitute precursors of disease to come on a long term basis (e.g., annoyance reactions, triggering of hypersensitivity reactions); (3) sensory warnings of exposure to harmful environmental factors (e-g., odour of toxic sulfides, mucosal irritation due to formaldehyde); (4) important tools in sensory bioassays for environmental characterisation (e.g.. using the odour criterion for general ventilation requirements or for screening of building materials to find those with low emissions of volatile organic compounds).

The senses responding to environmental exposure are not only hearing, vision, olfaction and taste, but also the skin and mucous membranes. As pointed out by WHO (1989), many different sensory systems that respond to irritants are situated on or near the body surface. Some of these systems tend to respond to an accumulated dose and their reactions are delayed. On the other hand, in the case of odor perception the reaction is immediate but also very much influenced by olfactory fatigue on prolonged exposures.

Responders are often unable to identify a single sensory system as the primary route of sensory irritation by airborne chemical compounds. The sensation of irritation is influenced by a number of factors such as previous exposures, skin temperature, competing sensory stimulation, etc. Since interaction and adaptation processes are characteristic of the sensory systems involved in the perception of odour and mucosal irritation, the duration of exposure influences the perception. Humans integrate different environmental signals to evaluate the total perceived air quality and assess comfort or discomfort. Comfort and discomfort by definition are psychological and for this reason the related symptoms, even when severe cannot be documented without using subjective reports. Sensory effects reported 10 be associated with IAP are in most cases multisensory and the same perceptions or sensations may originate from different sources. It is not known how different sensory perceptions are combined into perceived comfort and into the sensation of air quality. Perceived air quality is for example mainly related to stimulation of both the trigeminus and olfactorius nerves.

Several odorous compounds are also significant mucosal irritants, especially at high concentrations. The olfactory system signals the presence of odorous compounds in the air and has an important role as a warning system. In the absence of instrumentation for chemical detection of small amounts of some odorous vapours, the sense of smell remains the only sensitive indicator system. It is well known that environmental pollution can affect the nervous system. The effects of occupational exposure to organic solvents can be mentioned as an example. A wide spectrum of effects may be of importance, ranging from those at molecular level to behavioural abnormalities. Since the nerve cells of the CNS typically do not regenerate, toxic damage to them is usually irreversible. The nerve cells are highly vulnerable to any depletion in oxygen supply.

4.3.5. Cardiovascular Effects Associated With Iap

Increased mortality due to Cardiovascular Diseases (CVD) has been associated with exposure to ETS in some groups of non-smoking women married to smokers. Some investigators have also addressed the question whether total mortality is influenced by exposure to ETS, but results have been contradictory. As any effect on mortality would not be expected to occur until after many years of exposure, a problem in these types of study is the accuracy and reliability of the exposure classification. Attempts have also been made to relate ETS to electrocardiographic abnormalities and cardiovascular symptoms, but results have been inconclusive.

Carbon monoxide (CO) exerts its influence primarily through binding to the haemoglobin (Hb) in blood. The affinity of CO to Hb is about 200 times higher than the affinity of oxygen to Hb, so that at relatively low levels of CO in the air. Oxygen is replaced by CO. The percentage of Hb bound to CO (0/0 carboxyhaemoglobin) is a measure of recent exposure to CO. Organs with a high oxygen demand, such as the heart and the brain, are particularly susceptible to a reduced oxygenation caused by CO exposure. Early effects include reduction of time to onset of chest pain in exposed, exercising heart disease patients. At higher levels of exposure, myocardial infarctions may be triggered by CO.

5. BASIC CONTROL STRATEGIES

There are some basic control methods for lowering concentrations of indoor air pollutants (Ashford & Caldart, 2008), which are described bellow:

Source Management includes source removal, source substitution, and source encapsulation. Source management is the most effective control method when it can be practically applied. Source removal is very effective. However, policies and actions that keep potential pollutants from entering indoor are even better than preventing IAQ problems. Source substitution includes actions such as selecting a less toxic art material or interior paint than the products which are currently in use. Source encapsulation involves placing a barrier around the source so that it releases fewer pollutants into the indoor air (e.g., asbestos abatement, pressed wood cabinetry with sealed or laminated surfaces). Local Exhaust is very effecting on removing point sources of pollutants before they can disperse into the indoor air by exhausting the contaminated air outside. Well known examples include restrooms and kitchens where local exhaust is used. Other examples of pollutants that originate at specific points and that can be easily exhausted include science lab and housekeeping storage rooms, printing and duplicating rooms, and vocational/ industrial areas such as welding booths. Ventilation through use of cleaner (outdoor) air to dilute the polluted (indoor) air that people are breathing. Generally, local building codes specify the quantity (and sometimes quality) of outdoor air that must be continuously supplied to an occupied area. For situations such as painting, pesticide application, or chemical spills, temporarily increasing the ventilation can be useful in diluting the concentration of noxious fumes in the air. Exposure Control includes adjusting the time of use and location of use. An example of time of use for school students would be to strip and wax floors on Friday after school is dismissed, so that the floor products have a chance to off-gas over the location of use deals with moving the contaminating source as far as possible from occupants, or relocating susceptible occupants. Air Cleaning primarily involves the filtration of particles from the air as the air passes through the ventilation equipment. Gaseous contaminants can also be removed, but in most cases this type of system should be engineered on a case-by-case basis.

6. CONCLUSIONS

As it has been clearly proven above, indoor air pollution is a major public concern issue, which can be characterised as "global environmental phenomenon". Also, it is obvious, that the causes which created this domestic environmental problem, such as modern way of living, decreased ventilation rates for energy conservation or increased use of synthetic materials in buildings, are not expected to be reduced (it is more probable that they are going to be increased). Nevertheless, the task of reducing levels of exposure to air pollutants is rather complex. It begins with an analysis to determine which chemicals are present in the air, at what levels, and whether likely levels of exposure are hazardous to human health and the environment. It must then be decided whether an unacceptable risk is present. When a problem is identified, mitigation strategies have to be developed and implemented so as to prevent excessive risk to public health in the most efficient and cost-effective way. In addition, analyses of air pollution problems are exceedingly complicated. Some are national in scope (such as the definition of actual levels of exposure of the population, the determination of acceptable risk, and the identification of the most efficient control strategies), while others are of a more basic character and are applicable in all countries (such as analysis of the relationships between chemical exposure levels, and doses and their effects). So, it is very essential for governments of all countries- especially the governments of the more developed ones- to adopt and implement these policies, in order to effectively face this worldwide issue in combination with the need of energy saving, the use of new building materials and the modern trend of living.

REFERENCES

1 American Institute of Architects Research Corporation. 1979 Phase Two Report for the Development of Energy Performance Standards for New Buildings. Report to U.S. Department of Housing and urban Development and U.S. Department of Energy, 197

2 American Society of Heating, Refrigerating and Air-Conditioning engineers. 1972 ASHRAE Handbook of fundamentals. New York: American Society of Heating Refrigerating and Air-Conditioning Engineers, Inc., 688

3 D. J. Anderson, R. A. Hites, 1988 Chlorinated Pesticides in Indoor Air. Environ. Sci. Technol., 22 6 June 1988) 717720 , 0001-3936X

4 H. E. B. Andersson, Setterwall. A. . Klevard, Eds, 1996 The energy book. A resume of present knowledge and research, The Swedish council for building research, 9-15405-744-2 Sweden

5 M. Ando, K. Katagiri, K. Tamura, S. Yamamoto, M. Matsumoto, Y. F. Li, S. R. Cao, R. D. Ji, C. K. Liang, 1996 Indoor and Outdoor Air Pollution in Tokyo and Beijing Supercities. Atmos. Environ., 30 5 1996), 695702 , 1352-2310

6 N. A. Ashford, C. C. Caldart, 2008 Environmental Law, Policy, and Economics.Reclaiming the Environmental Agenda, 0-262-01238-3978-0-262-01238-6, MIT.

7 E. L. Avol, W. C. Navidi, S. D. Colome, 1998 Modeling Ozone Levels in and around Southern California Homes. Environ. Sci. Technol., 32 4 February 1998),463468 , 0001-3936X

8 S. Baek, O. , Y. Kim, S. , R. Perry, 1997 Indoor Air Quality in Homes, Offices, and Restaurants in Korean Urban Areas--Indoor/Outdoor Relationships. Atmos. Environ., 31 529544 , 1352-2310

9 A. Battaglia, D. Capra, G. Queirazza, A. Sampaolo, 1990 Radon exhalation rate in building materials and fly ashes in Italy, INDOOR AIR

'90, Proceedings of the 5th International Conference on Indoor Air Quality and Climate, 3 4752 . Toronto, 29 July- 3 August

10 F. Bochicchio, Venuti. G. Campos, F. Felici, A. Grisanti, G. Grisanti, F. Moroni, C. Nuccetelli, S. Risica, F. Tancredi, 1993 Characterisation of some parameters affecting the indoor radon exposure of the population, Proceedings of the First International Workshop on Indoor Radon Remedial Action, Rimini, Italy, 27th June- 2th July 1993

11 M. Brauer, P. B. Ryan, H. H. Suh, P. Koutrakis, J. D. Spengler, N. P. Leslie, I. H. Billick, 1990 Measurements of Nitrous Acid inside Two Research Houses. Environ. Sci. Technol., 24 10 October 1990), 15211527 , 0001-3936X

12 M. Brauer, J. D. Spengler, 1994 Nitrogen Dioxide Exposures inside Ice Skating Rinks. Am. J. Public Health, 84 3 March 1994), 429433 , 0090-0036

13 M. Brauer, K. Lee, J. D. Spengler, R. O. Salonen, A. Pennanen, O. A. Braathen, E. Mihalikova, P. Miskovic, A. Nozaki, T. Tsuzuki, S. Rui-Jin, Y. Xu, Z. Quing-Xiang, H. Drahonovska, S. Kjaergaard, 1997 Nitrogen Dioxide in Indoor Ice Skating Facilities: An International Survey. J. Air Waste Manage. Assoc., 47 (October 1997), 10951102 , 1047-3289

14 L. S. R. Brickus, J. N. Cardoso, Neto. F. R. De Quino, 1998 Distributions of Indoor and Outdoor Air Pollutants in Rio de Janeiro, Brazil: Implications to Indoor Air Quality in Bayside Offices. Environ. Sci. Technol, 32 22 November 1998) 34853490 , 0001-3936X

15 R. C. Brown, J. A. Hoskins, 1993 Asbestos and Man-made Mineral Fibres. MRC toxicology Unit, Woodmansterne Road, Surrey, 0-94841-107-4

16 S. K. Brown, M. R. Sim, M. J. Abramson, C. N. Gray, 1994 Concentrations of Volatile Organic Compounds in Indoor Air-A Review. Indoor Air, 4 2 June 1994), 123134 , 0905-6947

17 J. A. Bukowski, L. W. Meyer, 1995 Simulated Air Levels of Volatile Organic Compounds Following Different Methods of Indoor

Insecticide Application. Environ. Sci. Technol, 29 3 March 1995), 673676 , 0001-3936X

18 P. S. Burge, P. Jones, A. S. Robertson, 1990 Sick Building Syndrome: environmental comparisons of sick and healthy buildings. Indoor Air, 90 1, 479484

19 California Environmental Protection Agency, Office of Environmental. 1997 Health Hazard Assessment. Health Effects of Exposure to Environmental Tobacco Smoke. Final Report, September 1997.

20 Venuti. G. Campos, S. Colilli, A. Grisanti, G. Grisanti, G. Monteleone, S. Risica, G. Gobbi, M. P. Leogrande, A. Antonini, R. Borio, 1984 Indoor exposure in a Region of Central Italy, Radiat. Prot. Dosim, 7 1-41,271-274)

21 O. Castrén, A. Voutilainen, K. Winqist, I. Mäikeläinen, 1985 Studies of high indoor radon areas in Finland. Science of the Total Environment. 45 (October 1985), 311318 , 0048-9697

22 J. E. Cermak, 1976 Nature of air flow near buildings. ASHRAE Trans, 82 Pt.1), 10441054

23 C. Chan, C. , H. Ozkaynak, J. D. Spengler, L. Sheldon, 1991 Driver Exposure to Volatile Organic Compounds, CO, Ozone, and 2 under Different Driving Conditions. Environ. Sci. Technol., 25 No 6 (June 1991), 964972 , 0001-3936X

24 J. C. S. Chang, K. A. Krebs, 1992 Evaluation of Para-Dichlorobenzene Emissions from Solid Moth Repellant as a Source of Indoor Air Pollution. J. Air Waste Manage. Assoc., 42 9 September 1992), 12141217 , 1047-3289

25 F. S. Chapin, jr , 1974 Human activity patterns in the City: Things People Do in time and in space. New York: John Wiley & sons, Inc., 272 0-471-14563-7 13: 9780471145639

26 J. C. Chuang, G. A. Mack, M. R. Kuhlman, N. K. Wilson, 1991 Polycyclic Aromatic Hydrocarbons and Their Derivatives in Indoor and Outdoor Air in an Eight-Home Study. Atrnos. Environ., 25B 369380 , 1352-2310

27 K. D. Cliff, 1992 Thoron daughter concentrations in UK-homes, Radiat. Prot. Dosim., 45 1-4) Supplement, 361366

28 S. D. Colome, N. Y. Kado, P. Jaques, M. Kleinman, 1992 Indoor-Outdoor Air Pollution Relations: Particulate Matter Less Than 10 /zm in Aerodynamic Diameter (PM 10) in Homes of Asthmatics. Atmos. Environ., 26 1, 21732178 , 1352-2310

29 Committee on the Medical Effects of Air Pollutants: Guidance on the Effects on Health of Indoor Air Pollutants 2004 Department of Health, London

30 P. E. Converse, 1968 Time budgets. In D. L. Sills, Ed. International Encyclopedia of the Social sciences. 16 4247 New York: The Macmillan Company and the Free Press

31 D. R. Crump, D. Gardiner, 1989 Sources and Concentrations of Aldehydes and Ketones in Indoor Environments in the UK. Environ. Int., 15 455462 , 0160-4120

32 D. R. Crump, 1995 Volatile Organic Compounds in Indoor Air," in Issues in Environmental Science and Technology (R. E. Hester and R. M. Harrison, Eds.), Chap. 4, 109124 , Royal Chem. Soc., Letchworth, UK

33 J. M. Daisey, A. T. Hodgson, W. J. Fisk, M. J. Mendell, Brinke. J. Ten, 1994 Volatile Organic Compounds in Twelve California Office Buildings: Classes, Concentrations, and Sources. Atmos. Environ., 28 22 December 1994), 35573562 , 1352-2310

34 A. G. Davenport, 1960 A rationale for determination of design wirx: velocities. Proceedings of the ASCE Journal of the Structures Division, 86 3966

35 C. I. Davidson, S. Lin, F. , J. F. Osborn, M. R. Pandey, R. A. Rasmussen, M. A. K. Khalil, 1986 Indoor and Outdoor Air Pollution in the Himalayas. Environ. Sci. Technol., 20 6 June 1986), 561567 , 0001-3936X

36 R. J. De Meijer, P. Stoop, L. Put, 1992 Contribution of radon flows and radon sources to the radon concentration in a dwelling, Radiat. Prot. Dosim. 45 1-4) Supplement, 439442

37 T. Deyuan, 1993 Indoor and outdoor air radon concentration level in China", INDOOR AIR '93, Proceedings of the 6th International Conference on Indoor Air Quality and Climate, 4 459463 , Helsinki, Finland, July 4-81993

38 D. Dickerhoff, D. T. Grinsrud, B. Shohl, 1980 Infiltration and Air-Conditioning: A Case Study. Lawrence Berkeley Laboratory Report LBL-11674. Berkeley, Cal.: Lawrence Berkeley Laboratory

39 C. Dimitroulopoulou, S. R. Ashmore, M. T. R. Hill, M. A. Byrne, R. Kinnersley, 2006 INDAIR: A probabilistic model of indoor air pollution in UK homes. Atmos Environ, 40 63626379 , 1352-2310

40 M. Doi, S. Kobayashi, 1994 Characterization of Japanese wooden houses with enhanced radon and thoron concentrations, Health Phys., 66 3 274282 , 0017-9078

41 F. Dor, Y. Le Moullec, B. Festy, 1995 Exposure of City Residents to Carbon Monoxide and Monocyclic Aromatic Hydrocarbons during Commuting Trips in the Paris Metropolitan Area. J. Air Waste Manage. Assoc., 45 February 1995), 103110 , 1047-3289

42 G. Drakou, C. Zerefos, I. Ziomas, M. Voyatzaki, 1998 Measurements and Numerical Simulations of Indoor 0 3 and NO X in Two Different Cases. Atmos. Environ., 32 595610

43 A. V. Dremetsika, P. A. Siskos, E. B. Bakeas, 2005 Determination of formic and acetic acid in the interior atmosphere of display cases and cabinets in Athens Museums by reverse phase high performance liquid chromatography. Indoor Built. Environ., 14 1 5158

44 J. R. Druzik, M. S. Adams, C. Tiller, G. R. Cass, 1990 The Measurement and Model Predictions of Indoor Ozone Concentrations in Museums. Atmos. Environ., 24A 18131823 , 1352-2310

45 I. Ekinja, Z. Grabaric, B. S. Grabaric, 1995 Monitoring of Pyrocatechol Indoor Air Pollution. Atmos. Environ., 29 11651170 , 1352-2310

46 Collaborative. European, E. C. A. Action, 1998 Indoor air quality and its impact on man. Report 15 9-28270-119-0

47 European Collaborative Action (ECA). 1991 indoor air quality & its impact on man (formerly COST Project 61 3). Environment and Quality of Life Report 10 Effects of Indoor Air Pollution on Human Health

48 A. Febo, C. Perrino, 1991 Prediction and Experimental Evidence for High Air Concentration of Nitrous Acid in Indoor Environments. Atmos. Environ., 25A 10551061 , 1352-2310

49 A. Febo, C. Perrino, 1995 Measurement of High Concentration of Nitrous Acid inside Automobiles. Atmos. Environ., 29 345351 , 1352-2310

50 C. A. Federer, 1971 Effect of trees in modifying urban microclimate, pc, 2328 . In S. Little, and J. H. Noyes, Eds. Trees and Forests Urbanizing Environment. Amherst, Mass.: University of Massachusetts Cooperative Extension Service

51 S. E. Feinman, 1988 Formaldehyde Sensitivity and Toxicity, CRC Press, Boca Raton, FL

52 A. A. Fernandez-Bremauntz, M. R. Ashmore, 1995 Exposure of Commuters to Carbon Monoxide in Mexico City. Measurement of In-Vehicle Concentrations. Atmos. Environ., 29 525532 , 1352-2310

53 B. J. Finlayson-Pitts, J. N. Pitts, 1999 Chemistry of the upper and lower atmosphere Academic Press, 0-12-257060-x

54 M. L. Fischer, A. J. Bentley, K. A. Dunkin, A. T. Hodgson, W. W. Nazaroff, R. G. Sextro, J. M. Daisey, 1996 Factors Affecting Indoor Air Concentrations of Volatile Organic Compounds at a Site of Subsurface Gasoline Contamination. Environ. Sci. Technol., 30 10 September 1996), 29482957 , 0001-3936X

55 P. G. Flachsbart, G. A. Mack, J. E. Howes, C. E. Rodes, 1987 Carbon Monoxide Exposures of Washington Commuters. JAPCA, 37 2 135142 , 0894-0630

56 H. K. Florig, 1997 China's Air Pollution Risks. Environ. Sci. Technol., 31 274A279A

57 W. M. Foster, L. Jiang, P. T. Stetkiewicz, T. H. Risby, 1996 Breath Isoprene--Temporal Changes in Respiratory Output after Exposure to Ozone. J. Appl. Physiol., 80 706710

58 R. Geiger, 1965 The climate near the ground. Cambridge, Mass.: Harvard University Press, 611pp

59 D. Gelmont, R. A. Stein, J. F. Mead, 1981 Isoprene--The Main Hydrocarbon in Human Breath. Biochem. Biophys. Res. Commun., 99 4 April 1981), 14561460 , 0000-6291X

60 P. E. Georghiou, P. Blagden, D. A. Snow, L. Winsor, D. T. Williams, 1991 Mutagenicity of Indoor Air Containing Environmental Tobacco Smoke: Evaluation of a Portable PM-10 Impactor Sampler. Environ. Sci Technol., 25 8 August 1991), 14961500 , 0001-3936X

61 T. F. Gesell, 1983 Background atmospheric Rn-222 concentrations outdoors and indoors: a review, Health Phys., 45 289302 , 0017-9078

62 U. E. Gibson, R. E. Cawley, 1977 The heat pump solar collector interface-A practical experiment. Appliance Eng. 11 4), 6871

63 D. R. Gold, G. Allen, A. Damokosh, P. Serrano, C. Hayes, M. Castillejos, 1996 Comparison of Outdoor and Classroom Ozone

Exposures for School Children in Mexico City. J. Air Waste Manage. Assoc., P. B. 46 April 1996), 335342 , 1047-3289

64 R. L. Grasty, 1994 Summer outdoor radon variations in Canada and their relation to soil moisture, Health Phys. 66 2 185193 , 0017-9078

65 S. Gravesen, L. Larson, F. Gyntelberg, P. Skov, 1990 The role of potential immunogenic components of dust (MOD) in the sick building syndrome. Indoor Air, 90 1, 914 , 0905-6947

66 Q. Guo, M. Shimo, Y. Ikebe, S. Minato, 1992 The study of thoron and radon progeny concentrations in dwellings in Japan, Radiat. Prot. Dosim., 45 1-4) Supplement, 357359

67 C. H. Halios, C. G. Helmis, 2007 On the estimation of characteristic indoor air quality parameters using analytical and numerical methods. Science of the Total Environment, 381 222232

68 L. P. Hanrahan, H. A. Anderson, K. A. Daily, A. D. Eckmann, M. S. Kanarek, 1985 Formaldehyde Concentrations in Wisconsin Mobile Homes. JAPCA, 35 11641167 , 0894-0630

69 D. L. Heavner, M. W. Ogden, P. R. Nelson, 1992 Multisorbent Thermal Desorption/Gas Chromatography/Mass Selective Detection Method for the Determination of Target Volatile Organic Compounds in Indoor Air. Environ. Sci. Technol., 26 9 September 1992), 17371746 , 0001-3936X

70 C. G. Helmis, J. Tzoutzas, H. A. Flocas, V. Panis, O. I. Stathopoulou, V. D. Assimakopoulos, C. H. Halios, M. Apostolatou, G. Sgouros, E. Adam, 2007 Indoor Air Quality in a Dentistry Clinic. Science of the Total Environment, 377 349365

71 C. G. Helmis, V. D. Assimakopoulos, H. A. Flocas, O. I. Stathopoulou, G. Sgouros, M. Hatzaki, 2009 Indoor air quality assessment in the Air Traffic Control Tower of the Athens Airport, Greece. Environmental Monitoring and Assessment, 148 4760

72 M. W. M. Hisham, D. Grosjean, 1991 Sulfur Dioxide, Hydrogen Sulfide, Total Reduced Sulfur, Chlorinated Hydrocarbons, and Photochemical Oxidants in Southern California Museums. Atmos. Environ., 25A 14971505 , 1352-2310

73 A. T. Hodgson, J. M. Daisey, R. A. Grot, 1991 Sources and Source Strengths of Volatile Organic Compounds in a New Office Building. J. Air Waste Manage. Assoc., 41 11 November 1991), 14611468 , 1047-3289

74 A. T. Hodgson, K. Garbesi, R. G. Sextro, J. M. Daisey, 1992 Soil-Gas Contamination and Entry of Volatile Organic Compounds into a House near a Landfill. J. Air Waste Manage. Assoc., 42 3 March 1992), 277283 , 1047-3289

75 G. Hoek, B. Brunekreef, P. Hofschreuder, 1989 Indoor Exposure to Airborne Particles and Nitrogen Dioxide during an Air Pollution Episode. JAPCA, 39 13481349 , 0894-0630

76 R. D. Hopper, R. A. Levy, R. C. Rankin, M. A. Boyd, 1991 National ambient radon study, Proceedings of the International Symposium on Radon and Radon Reduction Technology, Philadelphia, PA (USA)

77 House of Commons Environment Committee. 1991 Indoor Pollution, 6th report, HMSO, London.

78 C. Howard, R. L. Corsi, 1998a Volatilization of Chemicals from Drinking Water to Indoor Air: The Role of Residential Washing Machines. J. Air Waste Manage. Assoc., 48 (October 1998), 907914 , 0000-1047- 3289

79 E. M. Howard, R. C. Mc Crillis, K. A. Krebs, R. Fortman, H. C. Lao, Z. Guo, 1998b Indoor Emissions from Conversion Varnishes. J. Air Waste Manage. Assoc., 48 (October 1998), 924930 , 0000-1047- 3289

80 G. Jakobi, P. Fabian, 1997 Indoor/Outdoor Concentrations of Ozone and Peroxyacetyl Nitrate (PAN). Int. J. Biometeorol., 40 3 May 1997), 162165 , 0020-7128

81 A. W. Jones, V. Lagesson, C. Tagesson, 1995 Origins of Breath Isoprene. J. Clin. Pathol., 48 10 October 1995), 979980 , 0021-9746

82 T. J. Kelly, 1996 Determination of Formaldehyde and Toluene Diisocyanate Emissions from Indoor Residential Sources. California Air Resources Board, Research Division, 2020 L Street, Sacramento, CA 95814, Final Report, Contract 93-315 93315 , November, 1996.

83 R. Kostiainen, 1995 Volatile Organic Compounds in the Indoor Air of Normal and Sick Houses. Atmos. Environ., 29 693702 , 1352-2310

84 P. A. Koushki, K. H. Al-Dhowalia, S. A. Niaizi, 1992 Vehicle Occupant Exposure to Carbon Monoxide. J. Air Waste Manage. Assoc., 42 12 December 1992), 16031608 , 1047-3289

85 P. Koutrakis, M. Brauer, S. L. K. Briggs, B. P. Leaderer, 1991 Indoor Exposures to Fine Aerosols and Acid Gases. Environ. Health Perspect., 95 2328

86 P. Koutrakis, S. L. K. Briggs, B. P. Leaderer, 1992 Source Apportionment of Indoor Aerosols in Suffolk and Onondaga Counties, New York. Environ. Sci. Technol., 26 3 March 1992), 521527 , 0001-3936X

87 K. W. Leovic, L. S. Sheldon, D. A. Whitaker, R. G. Hetes, J. A. Calcagni, J. N. Baskir, 1996 Measurement of Indoor Air Emissions from Dry-Process Photocopy Machines. J. Air Waste Manage. Assoc., 46 821829 , 1047-3289

88 J. I. Levy, K. Lee, J. D. Spengler, Y. Yanagisawa, 1998 Impact of Residential Nitrogen Dioxide Exposure on Personal Exposure: An International Study. J. Air Waste Manage. Assoc., 48 553560 , 1047-3289

89 C. W. Lewis, R. B. Zweidinger, 1992 Apportionment of Residential Indoor Aerosol VOC and Aldehyde Species to Indoor and Outdoor Sources, and Their Source Strengths. Atmos. Environ., 26A 21792184 , 1352-2310

90 J. Lewtas, S. Goto, K. Williams, J. C. Chuang, B. A. Petersen, N. K. Wilson, 1987 The Mutagenicity of Indoor Air Particles in a Residential Pilot Field Study: Application and Evaluation of New Methodologies. Atmos. Environ., 21 443449 , 1352-2310

91 G. Löfroth, P. I. Ling, E. Agurell, 1988 Public Exposure to Environmental Tobacco Smoke. Mutat. Res., 202 103110 , 0027-5107

92 M. P. Ligocki, L. G. Salmon, T. Fall, M. C. Jones, W. W. Nazaroff, G. R. Cass, 1993 Characteristics of Airborne Particles inside Southern California Museums. Atmos. Environ., 27A 697711 , 1352-2310

93 L. Liu, J. S. , P. Koutrakis, H. H. Suh, J. D. Mulik, R. M. Burton, 1993 Use of Personal Measurements for Ozone Exposure Assessment: A Pilot Study. Environ. Health Perspect., 101 318324

94 G. Löfroth, C. Stensman, M. Brandhorst-Satzkorn, 1991 Indoor Sources of Mutagenic Aerosol Particulate Matter: Smoking, Cooking, and Incense Burning. Mutat. Res., 261 2128 , 0000-0027- 5107

95 M. C. Marbury, D. P. Harlos, J. M. Samet, J. D. Spengler, 1988 Indoor Residential 2 Concentrations in Albuquerque, New Mexico. JAPCA, 38 392398 , 0894-0630

96 T. Mc Curdy, G. Glen, L. Smith, 2000 The National Exposure Research Laboratory's Consolidated Human Activity Database. J Exposure Analysis Environ Epidemiol, 10 6 November 2000), 566578 , 1559-0631

97 T. E. Mc Kone, 1987 Human Exposure to Volatile Organic Compounds in Household Tap Water: The Indoor Inhalation Pathway. Environ. Sci. Technol., 21 12 December 1987), 11941201 , 0001-3936X

98 D. A. Mclntyre, 1980 Indoor Climate. London: Applied Science Publishers Ltd. 443 pp, 0-85334-868-5

99 S. Mendis, P. A. Sobotka, D. E. Evler, 1994 Pentane and Isoprene in Expired Air from Humans--Gas Chromatographic Analysis of a Single Breath. Clin. Chem., 40 8 August 1994), 14851488 , 0009-9147

100 B. Meyer, K. Hermanns, 1986 Formaldehyde Release from Wood Products. An Overview. in Formaldehyde Release from Wood Products, ACS Symposium Series 316, Chapter 1, 116 , Am. Chem. Soc., Washington, DC, 9780841209824

101 W. Michelson, 1973 Time-budgets in environmental research: Some introductory considerations, 262268 . In W. F. E. Preiser, Ed. Environmental Design Research. 2 symposia and Workshops. Fourth International EDRA Conference. Stroudsburg, Pa.: Dowden, Hutchinson, and Ross, Inc.

102 J. Miller, 1990 Fungi as contaminants in indoor air. Indoor Environment, 90 5, 5164

103 S. Mitra, B. Ray, 1995 Patterns and Sources of Polycyclic Aromatic Hydrocarbons and Their Derivatives in Indoor Air. Atmos. Environ., 29 33453356 , 1352-2310

104 A. Morey, 1990 Biological contaminants in the Indoor Environment ASTM. ISBN-EB: 978-0-8031-5131-4

105 D. J. Moschandreas, S. S. Morse, 1979 The relationship between energy conservation measures and exposure to indoor toxic pollutants. Paper 27 b., presented at the American Institute of Chemical Engineers 87th National Meeting, Boston, Massachusetts, August 19-22, 1979

106 D. J. Moschandreas, Ed, 1978 Indoor Air pollution in the Residential Environment. Vol. II. Field Monitoring Protocol, Indoor Episodic Pollutant Release Experiments and Numerical Analyses. U.S. Environmental Protection Agency Report EPA-600 -7-78-229b. Research triangle Park: U.S. Environmental Protection Agency, Environmental Monitoring and Support Laboratory, 1978, 240 pp.

107 D. J. Moschandreas, J. Zabransky, D. J. Pelton, 1980 Comparison of Indoor-Outdoor concentrations of Atmospheric Pollutants. GEOMET Report ES-823 . Palo Alto, Cal.: Electric Power Research Institute

108 D. J. Moschandreas, D. S. O'Dea, 1995 Measurement of Perchloroethylene Indoor Air Levels Caused by Fugitive Emissions from

Unvented Dry-to-Dry Dry Cleaning Units. J. Air Waste Manage. Assoc., 45 February 1995), 111115 , 1047-3289

109 C. L. Moseley, M. R. Meyer, 1992 Petroleum Contamination of an Elementary School: A Case History Involving Air, Soil-Gas, and Groundwater Monitoring. Environ. Sci. Technol., 26 1 January 1992), 185192 , 0001-3936X

110 J. L. Mumford, X. Z. He, R. S. Chapman, S. R. Cao, D. B. Harris, X. M. Li, Y. L. Xian, W. Z. Jiang, C. W. Xu, J. C. Chuang, W. E. Wilson, M. Cooke, 1987 Lung Cancer and Indoor Air Pollution in Xuan Wei, China. Science, 235 4785 January 1987), 217220 , 0036-8075

111 J. L. Mumford, R. W. Williams, D. B. Walsh, R. M. Burton, D. J. Svendsgaard, J. C. Chuang, V. S. Houk, J. Lewtas, 1990 Indoor Air Pollutants from Unvented Paraffin Heater Emissions in Leaderer, B. P., P. M. Boone, and S. K. Hammond, "Total Particle, Sulfate, and Acidic Aerosol Emissions from Paraffin Space Heaters. Environ. Sci. Technol., 24 6 June 1990), 908912 , 0001-3936X

112 L. M. Neas, D. W. Dockery, J. H. Ware, J. D. Spengler, B. G. Ferris, Jr , F. E. Speizer, 1994 Concentration of Indoor Particulate Matter as a Determinant of Respiratory Health in Children. Am. J. Epidemiol., 139 11 June 1994), 10881099 , 0002-9262

113 P. R. Nelson, S. P. Kelly, F. W. Conrad, 1998 Studies of Environmental Tobacco Smoke Generated by Different Cigarettes. J. Air Waste Manage. Assoc., 48 April 1998), 336344 , 1047-3289

114 NENOECD (Nuclear Energy Agency Organisation for Economic Co-operation and Development) 1979 Exposure to Radiation from the Natural Radioactivity in Building Materials, Expert Group Report, OECD, Paris

115 A. V. Nero, Jr (1988, In, (1988).In:Radon and its decay products in indoor air: an overview, In Radon and its decay products in Indoor Air, W.W. Nazaroff and A.V. Nero Jr. edit., 156 156 Wiley Interscience, 0-47162-810-7

116 A. V. Nero, Jr , 1989 Earth, Air, Radon and Home. Physics Today. 42 4 (April 1989), 3239 , 0031-9228

117 F. J. Offermann, S. A. Loiselle, A. T. Hodgson, L. A. Gundel, J. M. Daisey, 1991 A Pilot Study to Measure Indoor Concentrations and Emission Rates of Polycyclic Aromatic Hydrocarbons. IndoorAir, 4 497512 , 0905-6947

118 W. Ott, P. Switzer, N. Willits, Carbon Monoxide Exposures inside an Automobile Traveling on an Urban Arterial Highway. 1994 J. Air Waste Manage. Assoc., 44 August 1994), 10101018 , 1047-3289

119 W. R. Ott, 1995 Human activity patterns: A review of the literature for air pollution exposure estimation. SIMS Technical report. Stanford Cal.: University, Department of statistics

120 J. R. Ottensman, 1972 Systems of Urban Activities and Time: An interpretive Review of the Literature. An Urban Studies Research Paper. Chapel Hill, N.C.: University of North Carolina, Center for Urban and Regional Studies, 1972. 45pp.

121 A. S. Pennanen, R. O. Salonen, S. Aim, M. J. Jantunen, P. Pasanen, 1997 Characterization of Air Quality Problems in Five Finnish Indoor Ice Arenas. J. Air Waste Manage. Assoc., 47 October 1997), 10791086 , 1047-3289

122 R. Perry, P. W. Kirk, 1986 Indoor and Ambient Air Quality. Selper, London

123 R. Perry, I. L. Gee, 1994 Vehicle Emissions and Effects on Air Quality: Indoors and Outdoors. Indoor Environ., 3 224236

124 J. A. Peterka, J. E. Cermak, 1977 Turbulence in Building wakes, 447463 . In K.J. Eaton, Ed. Proceedings of the fourth international conference on wind effects on buildings and structures, Heathrow, 1975. New York: Cambridge University Press, 0-52120-801-7

125 M. Petreas, K. Liu, S. , B. Chang, H. , S. B. Hayward, K. Sexton, 1988 A Survey of Nitrogen Dioxide Levels Measured inside Mobile Homes. JAPCA, 38 647651 , 0894-0630

126 M. Phillips, J. Greenberg, J. Awad, 1994 Metabolic and Environmental Origins of Volatile Organic Compounds in Breath. J. Clin. Pathol., 47 11 November 1994), 10521053 , 0021-9746

127 J. N. Pitts, T. J. Wallington, H. W. Biermann, A. M. Winer, 1985 Identification and Measurement of Nitrous Acid in an Indoor Environment. Atmos. Environ., I9 763767 , 1352-2310

128 J. N. Pitts, H. W. Biermann, E. C. Tuazon, M. Green, W. D. Long, A. M. Winer, 1989 Time-Resolved Identification and Measurement of Indoor Air Pollutants by Spectroscopic Techniques: Gaseous Nitrous Acid, Methanol, Formaldehyde, and Formic Acid. JAPCA, 39 13441347 , 0894-0630

129 J. G. Price, J. G. Rigby, L. Christensen, R. Hess, D. D. La Pointe, A. R. Ramelli, M. Desilets, R. D. Hopper, T. Kluesner, S. Marshall, 1994 Radon in outdoor air in Nevada, Health Phys. 66 4 433438 , 0017-9078

130 Priorities for Indoor Air Research and Action. 1991 International Conference, Montreux.

131 J. J. Quackenboss, M. D. Lebowitz, C. D. Crutchfield, 1989 Indoor-Outdoor Relationships for Particulate Matter: Exposure Classifications and Health Effects. Environ. Int., I5 353360 , 0160-4120

132 R. Reiss, P. B. Ryan, P. Koutrakis, 1994 Modeling Ozone Deposition onto Indoor Residential Surfaces. Environ. Sci. Technol., 28 3 March 1994), 504513 , 0001-3936X

133 R. Reiss, P. B. Ryan, S. J. Tibbetts, P. Koutrakis, 1995a Measurement of Organic Acids, Aldehydes, and Ketones in Residential Environments and Their Relation to Ozone. J. Air Waste Manage. Assoc., 45 October 1995), 811822 , 1047-3289

134 R. Reiss, P. B. Ryan, P. Koutrakis, S. J. Tibbets, 1995b Ozone Reactive Chemistry on Interior Latex Paint. Environ. Sci. Technol., 29 8 August 1995), 19061912 , 0001-3936X

135 E. M. C. Rob, A. Rannou, J. Le Bronec, 1992 Radon measurement in the environment in France, Radiat. Prot. Dosim. 45 1-4) Supplement, 455457

136 I. Romieu, M. C. Lugo, S. Colome, A. M. Garcia, M. H. Avila, A. Geyh, S. R. Velasco, E. P. Rendon, 1998 Evaluation of Indoor Ozone Concentration and Predictors of Indoor-Outdoor Ratio in Mexico City. J. Air Waste Manage. Assoc., 48 April 1998), 327335 , 1047-3289

137 D. R. Rowe, A. , K. H. Dhowalia, M. E. Mansour, 1989 Indoor-Outdoor Carbon Monoxide Concentrations at Four Sites in Riyadh, Saudi Arabia. J. Air Waste Manage. Assoc., 39 8 August 1989), 11001102 , 1047-3289

138 D. R. Rowe, K. H. Al-Dhowalia, M. E. Mansour, 1991 Indoor-Outdoor Nitric Oxide and Nitrogen Dioxide Concentrations at Three Sites in Riyadh, Saudi Arabia. J. Air Waste Manage. Assoc., 41 7 July 1991), 973976 , 1047-3289

139 P. B. Ryan, M. L. Soczek, J. D. Spengler, I. H. Billick, 1988 The Boston Residential 2 Characterization Study: I. Preliminary Evaluation of the Survey Methodology. JAPCA, 38 2227 , 0894-0630

140 L. G. Salmon, W. W. Nazaroff, M. P. Ligocki, M. C. Jones, G. R. Cass, 1990 Nitric Acid Concentrations in Southern California Museums. Environ. Sci. Technol., 24 7 July 1990), 10041012 , 0001-3936X

141 M. Santamouris, N. Papanikolaou, I. Livada, I. Koronakis, C. Georgakis, A. Argiriou, D. Assimakopoulos, 2001 On the impact of Urban climate on the energy consumption of buildings. Solar Energy, 70 201216

142 V. H. Schaeffer, B. Bhooshan, S. Chen, B. , J. S. Sonenthal, A. T. Hodgson, 1996 Characterization of Volatile Organic Chemical Emissions

from Carpet Cushions. J. Air Waste Manage. Assoc., 46 September 1996), 813820 , 1047-3289

143 G. Sciocchetti, G. F. Clemente, G. Ingrao, F. Scacco, 1983 Results of a survey on radioactivity of building materials in Italy. Health Phys. 45 2 August 1983), 385388 0017-9078

144 G. Sciocchetti, M. Bovi, G. Cotellessa, P. G. Baldassini, C. Battella, I. Porcu, 1992 Indoor radon and thoron surveys in high radioactivity areas of Italy, Radiat. Prot. Dosim., 45 1-4) Supplement, 509513

145 K. Sexton, M. X. Petreas, K. S. Liu, Formaldehyde Exposures inside Mobile Homes 1989 Environ. Sci. Technol., 23 8 August 1989), 985988 , 0001-3936X

146 J. J. Shah, H. B. Singh, 1988 Distribution of Volatile Organic Chemicals in Outdoor and Indoor Air. Environ. Sci. Technol., 22 11 November 1988), 13811388 , 0001-3936X

147 F. H. Shair, K. L. Heitner, 1974 Theoretical Model for relating indoor pollution concentrations to those outside. Environ. Sci. Technol, 8 5 May 1974), 444451 , 0001-3936X

148 J. L. Shepherd, R. L. Corsi, J. Kemp, 1996 Chloroform in Indoor Air and Wastewater: The Role of Residential Washing Machines. J. Air Waste Manage. Assoc., 46 July 1996), 631642 , 1047-3289

149 M. H. Sherman, 1981 Air Infiltration in Buildings. Berkeley, Cal.: University of California, Ph.D. Dissertation,. (published as Lawrence Berkeley Laboratory Report LBL-1U712)

150 H. C. Shields, C. J. Weschler, (1992, (1992).Volatile Organic Compounds Measured at a Telephone Switching Center from 5/30/85-12/6/88: A Detailed Case Study. J. Air Waste Manage. Assoc., 42 42 6 June 1992), 792804 , 1047-3289

151 H. C. Shields, D. M. Fleischer, C. J. Weschler, 1996 Comparisons among VOCs Measured in Three Types of U.S. Commercial Buildings with Different Occupant Densities," Indoor Air, 6 217 , 0905-6947

152 N. Shinohara, Y. Kai, A. Mizukoshi, M. Fujii, K. Kumagai, Y. Okuizumi, M. Jona, Y. Yanagisawa, 2009 On-site passive flux sampler measurement of emission rates of carbonyls and VOCs from multiple indoor sources. Building Environ, 44 859863 , 0360-1323

153 S. Short, 2001 The issues and implications of setting and applying Indoor Air Quality Guidelines. IEH Seminar, University of Leicester,http://www.le.ac.uk/ieh/pdf/IAQmtgrep.

154 P. A. Siskos, K. E. Bouba, A. P. Stroubou, 2001 Determination of selected pollutsnts and measurements of physical parameters of the evaluation of indoor air quality in school buildings in Athens, Greece. Indoor Built. Environ., 10 185192

155 P. A. Siskos, 2003 An experimental Investigation of the Indoor Air Quality in Fifteen School Buildings in Athens, Greece. International Journal of Ventilation, 2 3 185201 , 1473-3315

156 A. P. Soldatos, E. B. Bakeas, P. A. Siskos, 2003 Occupational exposure to btex of workers in car parkings and gasoline service stations in Athens, Greece. Fresenius Environmental Bulletin, 12 9 10641070

157 S. Sollinger, K. Levsen, G. Wfinsch, 1993 Indoor Air Pollution by Organic Emissions from Textile Floor Coverings. Climate Chamber Studies under Dynamic Conditions. Atrnos. Environ., 27B 183192 , 1352-2310

158 S. Sollinger, K. Levsen, G. Wiinsch, 1994 Indoor Pollution by Organic Emissions from Textile Floor Coverings: Climate Test Chamber Studies under Static Conditions. Atrnos. Environ., 28 23692378 , 1352-2310

159 J. D. Spengler, D. W. Dockery, W. A. Turner, J. M. Wolfson, B. G. Ferris, 1981 Long-Term Measurements of Respirable Sulfates and Particles inside and outside Homes. Atrnos. Environ., 15 2330 , 1352-2310

160 J. D. Spengler, M. Brauer, J. M. Samet, W. E. Lambert, 1993 Nitrous Acid in Albuquerque, New Mexico, Homes. Environ. Sci. Technol., 27 5 May 1993), 841845 , 0001-3936X

161 J. Spengler, M. Schwab, P. B. Ryan, S. Colome, A. L. Wilson, I. Billick, E. Becker, 1994 Personal Exposure to Nitrogen Dioxide in the Los Angeles Basin. J. Air Waste Manage. Assoc., 44 January 1994), 3947 , 1047-3289

162 C. W. Spicer, D. V. Kenny, G. F. Ward, I. H. Billick, 1993 Transformations, Lifetimes, and Sources of 2 HONO, and HNO 3 in Indoor Environments. J. Air Waste Manage. Assoc., 43 November 1993), 14791485 , 1047-3289

163 T. H. Stock, 1987 Formaldehyde Concentrations inside Conventional Housing. JAPCA, 37 913918 , 0894-0630

164 G. A. Swedjemark, L. Mjönes, 1984 Radon and radon daughter concentration in Swedish Homes, Radiat. Prot. Dosim. 7 1-4), 341345

165 G. T. Tamura, 1975 Measurement of air leakage characteristics of house enclosures. ASHRAE Trans, 81 Pt. 1), 202211

166 C. V. Thompson, R. A. Jenkins, C. E. Higgins, 1989 A Thermal Desorption Method for the Determination of Nicotine in Indoor Environments. Environ. Sci. Technol., 23 4 April 1989), 429435 , 0001-3936X

167 B. A. Tichenor, L. A. Sparks, J. B. White, M. D. Jackson, 1990 Evaluating Sources of Indoor Air Pollution. J. Air Waste Manage. Assoc., 40 4 April 1990), 487492 , 1047-3289

168 G. W. Traynor, M. G. Apte, H. A. Sokil, J. C. Chuang, W. G. Tucker, J. L. Mumford, 1990 Selected Organic Pollutant Emissions from Unvented Paraffin Space Heaters. Environ. Sci. Technol., 24 9 September 1990), 12651270 , 0001-3936X

169 M. Tsakas, P. Siskos, 2010 Indoor Air Quality in the Control Tower of Athens International Airport, Greece. Indoor and Built Environment, 0 0106 , 0142-0326X

170 J. L. Tyson, P. W. Fairey, C. R. Withers, 1993 Elevated radon levels in ambient air, INDOOR AIR '93, Proceedings of the 6th International Conference on Indoor Air Quality and Climate,, 1 4, 443448 . Helsinki, Finland, July 4-8,1993

171 UNSCEAR (United Nations Scientific Committee on the Effects of Atomic Radiation) 1977 Sources and effects of ionizing radiation, United Nations ed., New York, E.77.IX.1.

172 UNSCEAR (United Nations Scientific Committee on the Effects of Atomic Radiation) 1993 Sources and effects of ionizing radiation, United Nations ed., New York, E.94.IX.2.

173 W. A. Wadc, W. A. Cote, J. E. Yocom, 1975 A Study of Indoor Air Quality. JAPCA, 25 933939 , 0894-0630

174 J. Waldman, P. Jenkins, 2004 Indoor air exposure research and reducing indoor air exposures in California. In: Krzyzanowski, M., Jantunen, M., Bartonova, A., Oglesby, L., Kepahalopoulos, S., Kotzias D, (Eds.). Role of Human assessment in air quality management. Report EUR 21052, European Communities, Luxembourg, 6671

175 J. C. Wallace, L. P. Brzuzy, S. L. Simonich, S. M. Visscher, R. A. Hites, 1996 Case Study of Organochlorine Pesticides in the Indoor Air of a Home. Environ. Sci. Technol., 30 9 August 1996), 27152718 , 0001-3936X

176 L. Wallace, 1996 Indoor Particles: A Review. J. Air Waste Manage. Assoc., 46 February 1996), 98126 , 1047-3289

177 N. V. Waubke, W. Kusterle, 1990 Mould infestations in residential buildings. Paper presented at the 1st symposium on mould infestations, Innsbruck, January 1990

178 C. J. Weschler, H. C. Shields, D. V. Naik, 1989 Indoor Ozone Exposures. JAPCA, 39 15621568 , 0894-0630

179 C. J. Weschler, H. C. Shields, D. Rainer, 1990 Concentrations of Volatile Organic Compounds at a Building with Health and Comfort Complaints. Am. Ind. Hyg. Assoc. J., 51 261268 , 0002-8894

180 C. J. Weschler, H. C. Shields, D. V. Naik, 1994 Indoor Chemistry Involving O3, NO, and 2 as Evidence by 14 Months of Measurements at a Site in Southern California. Environ. Sci. Technol., 28 No. 12, 21202132 , 0001-3936X

181 C. J. Weschler, H. C. Shields, 1996 Chemical Transformations of Indoor Air Pollutants, in Indoor Air '96 (S. Yoshizawa, K. Kimura, K. Ikeda, S. Tanabe, and T. Iwata, Eds.). Organizing Committee of the 7th International Conference of Indoor Air Quality and Climate, 1 919924 , Nagoya, Japan

182 D. L. West, 1977 Contaminant dispersion and dilution in a ventilated space. ASHRAE Trans, 83 Pt. 1), 125140

183 R. F. White, 1995 Landscape development and natural ventilation. Effect of moving air on buildings and adjacent areas. Landscape Archit, 45 7281 , 0023-8031

184 P. Wolkoff, 1999 Photocopiers and Indoor Air Pollution. Atmos. Environ., 33 21292130 , 1352-2310

185 World Health Organization. 1986 Regional Seminar on Human Exposure to Outdoor and Indoor Air Pollution and its Effects on Health. PEPAS, Kuala Lumpur

186 World Health Organization 1989 Indoor air quality: Organic Pollutants. Copenhagen, EURO 111

187 J. Zhang, Q. He, P. J. Lioy, 1994 Characteristics of Aldehydes: Concentrations, Sources, and Exposures for Indoor and Outdoor Residential Microenvironments. Environ. Sci. Technol., 28 1 January 1994), 146152 , 0001-3936X

188 L. Z. Zhang, J. L. Niu, 2004 Modeling VOCs emissions in a room with a single-zone multi-component multi-layer technique. Building Environ, 39 523531 , 0360-1323

CHAPTER 5

Indoor Air Quality - Volatile Organic Compounds: Sources, Sampling and Analysis

Alessandro Bacaloni, Susanna Insogna and Lelio Zoccolillo

[1] Department of Chemistry, University of Rome "La Sapienza", Italy

1. INTRODUCTION

Since the 70s, research has found in Europe and in the United States that individuals spend between 70 and 90% of their time indoors. Health studies have found that exposures to a variety of air pollutants indoors can be substantially higher than outdoors, even in urban environment. Volatile Organic Compounds (VOCs) are often considered among the more important indoor pollutants, because of by their continue emission from many sources and their diffusion properties. With the aim to evaluate the occupants' discomfort and health effects and in order to develop guidelines and standards, Indoor Air Quality (IAQ) assessment and control is an essential step. IAQ assessment will complain:

- Sources: Identification and characterization of sources, as emissions from materials, products or activities, is best done under laboratory conditions; so it is possible measuring rates of emissions (especially chemicals such as VOCs). Exposure characterization is the second level of source identification; after the measurement of contaminants' concentrations in controlled environment, characterised by known sources, adsorbing and absorbing surfaces, these data can be used in validation of current exposure models.

- Sampling methods: In order to determine concentrations of VOCs and exposures of building occupants via inhalation, field studies can be carried out by sampling methods (and analysis) in accordance with existing official methods (EC, NIOSH, OSHA, ACGIH, etc.). This way may be expensive and cumbersome; in addition, it can be not exhaustive in predicting the discomfort and health impact. A greater number of perspectives are offered by using some "descriptors" that can be more adequate in characterizing anthropogenic pollution. Specific sampling methods may be reserved for contaminants with specific toxic effects (e.g., formaldehyde, benzene, monomers, etc.). Measurements of specific contaminants' can be necessary for sources that cause high room concentrations for relatively short periods (e.g., in case of freshly applied coatings on walls, etc.). Currently, diffusion (passive) samplers are mainly used in order to evaluate long term exposures (days to weeks and more)

- Analysis and data meaning: existing analytical methods are validated and generally show adequate limits of detection (LODs) and limits of quantification (LOQs) even in measuring subtoxic contaminants' concentrations. Analytical methods for the more frequent indoor contaminants are presented and commented. More difficulties lie in the interpretation of results, due to the limited indications suggested by European and International Standards (such as EN, EN ISO, etc.) that will be discussed. The difficulty in data evaluation is growing depending on the simultaneous occurrence of contaminants at subtoxic concentration.

In this chapter, there will be presented the way to handle the problem in IAQ assessment, and some practical applications, in order to provide the logical pathway to face the majority of actual cases.

2. INDOOR AIR QUALITY AND VOCS

Research on pollutant sources is needed to identify pollutants and emission levels from building materials and other products. Work in this area will serve two purposes: (1) providing measurement protocols and data to employ in exposure reduction actions, and (2) providing correlations between research on health effects and pollutant sources of contaminant critical levels. Finally, research is required to determine definition, causes of, and solutions to multiple chemical sensitivity (MCS). Identifying the physiologic nature of MCS is the first step in understanding whether and how IAQ eventually contributes to the syndrome. Maroni et al. (1995) recommended the following definitions that should be used in the description of indoor related complaints and illnesses:

- Building-Related Environmental Complaints (BREC): Complaints of poor IAQ or poor indoor air environment. BREC are usually registered in the complaint (annoyance) part of questionnaire studies.

- Building-Related Symptoms (BRS) or Building Related Health Complaints (BRC): The health complaints (subjective symptoms) reported by the single individual as occurring inside a building and usually subsiding shortly after leaving it.

- Sick Building Syndrome (SBS): denotes a situation where a significant number of the occupants of a building complain of a typical group of general, unspecific, and irritating symptoms, including particularly headache, lethargy, dry eyes, blocked nose, and sore throat. The symptoms usually fade after the person has left the indoor environment but the specific cause is unidentified.

- Building-Related Illness (BRI): clinical condition with defined symptoms and signs in which the cause (aetiology) is building-related and identifiable.

The difference between BRI and SBS is that the building problems are identified in the former and undiagnosed in the latter. The term "sick building" probably should not be used on its own, but it might be replaced with the expression "building with indoor climate problems" or "problem building."

VOCs are often more important in assessing IAQ because of their ubiquity; VOCs may be used like "descriptors" that can be more adequate in characterizing anthropogenic pollution. Specific sampling methods may be reserved for contaminants with specific toxic effects (e.g., formaldehyde, benzene, monomers, etc.). In many cases, investigations of the indoor air are carried out because of complaints about poor IAQ, which are made by persons living or working indoors. Such complaints may be perception of unknown or unpleasant odours, headache, irritation of the eyes, nose and throat, dryness of the skin or symptoms like tiredness, lack of concentration and unspecific hypersensitivity reactions. Investigations resulting from observed or suspected health problems of occupants are quite similar and require the same sampling strategy.

3. SOURCES

Identification of indoor pollutant sources may be very important; emission from point (or surface) sources generates a pronounced concentration gradient near the source. The existence of such concentration gradients can be used to trace the source in an indoor environment by a careful selection of sampling sites. This strategy can lead to recommendations on how to improve indoor air quality, e.g. removing identified materials and strong emitters.

Maroni et al (1995) proposed these recommendations concerning source control:

Material Selection

- Manufacturers should be required to provide sufficient information for the evaluation of their products' safety. Protection of legitimate confidentiality must not impede evaluation.
- Standardization and harmonization of emission testing procedures must continue at international level. Meanwhile, research and application of different techniques and methodological intercomparison should be encouraged.
- Labelling of products and their ranking for safety and emission properties should be encouraged to facilitate appropriate selection and use by consumers. The choice of process may vary according to the nature of the risk involved and the local priorities.

Biological Contaminants

- The growth of moulds and other microorganisms in the indoor environment should be avoided. This can be achieved effectively by elimination of moisture.
- Good housekeeping and sanitation rather than the use of biocides are appropriate control measures also for mites.

Volatile Organic Compounds (VOCs)

- Overall exposure to VOCs should be kept at the lowest possible level primarily by proper control of the emission sources.
- Pending improved methods, the total volatile organic compounds concept is a useful tool for practical assessment of the overall quality of materials and of indoor air. However, at least the type of major VOC species must be known.
- More research is recommended for the sensory effects of VOC mixtures and for some individual compounds typically present in

indoor air mixtures for which little toxicological knowledge is available; this recommendation is strongly current, due to the growing interest in persistent organic pollutants (POPs) as endocrine disruptors, also in classic subtoxic concentrations.

4. SAMPLING METHODS

The primary objective of taking indoor air is to determine the quality of indoor air with the aim of assessing any risk to the health of the population and of individuals due to indoor air pollution. Monitoring is useful also in testing effectiveness of remedial actions, such as modifications to a building, its systems or equipment, aimed at reducing indoor air pollution. The measurements for testing their effectiveness are comparison measurements before and after the remedial actions. It is also important to record intermediate and long-term trends of indoor air pollution concentration. Respective analysis of trends will help to maintain, improve and establish abatement or risk management procedures. Finally, beside screening measurements, it can be very helpful to evaluate indoor air concentrations at abnormal or "worst-case" conditions of the indoor climate (temperature, humidity, ventilation etc.) and during particular activities. Sampling and analytical methods may be also used in validation of indoor pollution models.

To determine reference values of indoor air pollutants, the following elements of sampling strategy shall be used in addition to the basic information characterising every sampling method (EN 14412, 2004):

- Time of sampling: to rule out any seasonal effects the individual sampling events shall be evenly distributed over the year. The time of individual sampling shall be fixed in a way that representative concentration values can be assessed. However, for sampling for ≥ one week, the choice of sampling time tends to be less important.

- Sampling duration and sampling frequency: the duration of sampling has to be set in a way that the limits of determination for all compounds of interest are exceeded and that representative readings are obtained. It is necessary to maintain a sampling duration of one week or a multiple to achieve a maximum of representativity per sampling site. This is to assess all different pollution loads that occur on the different days of the week at least once. One sampling per site is adequate.

- Sampling site and determinations with resolution in space. The sampling site shall be representative, e.g. the centre of the room most inhabited by all occupants shall be used.

General methods (like NIOSH, OSHA or similar) developed for industrial workplaces may be adapted for VOC determination in air; as an example, we report the synthesis of NIOSH Method 2549 (Table 1).

APPLICABILITY: This method has been used for the characterization of environments containing mixtures of volatile organic compounds (see Table 2). The sampling has been conducted using multi-bed thermal desorption tubes. The analysis procedure has been able to identify a wide range of organic compounds, based on operator expertise and library searching.

INTERFERENCES: Compounds which coelute on the chromatographic column may present an interference in the identification of each compound. By appropriate use of background subtraction, the mass spectrometrist may be able to obtain more representative spectra of each compound and provide a tentative identity (see Table 2).

OTHER METHODS: Other methods have been published for the determination of specific compounds in air by thermal desorption/gas chromatography [1-3]. One of the primary differences in these methods is the sorbents used in the thermal desorption tubes.

REAGENTS:

- Air, dry
- Helium, high purity

- Organic compounds of interest for mass spectra verification (See Table 2).*
- Solvents for preparing spiking solutions: carbon disulfide (low benzene chromatographic grade), methanol, etc. (99+% purity)

EQUIPMENT:

- Sampler: Thermal sampling tube, ¼" s.s. tube, multi-bed sorbents capable of trapping organic compounds in the C_3-C_{16} range. Exact sampler configuration depends on thermal desorber system used. See Figure 1 as an example.

Table 1. NIOSH method 2549: volatile organic compounds (modified).

SAMPLING AND MEASUREMENT	
SAMPLER:	THERMAL DESORPTION TUBE (multi-bed sorbent tubes containing graphitized carbons and carbon molecular sieve sorbents [See Appendix])
FLOW RATE:	0.01 to 0.05 L/min
VOL -MIN:	1L
VOL -MAX:	6L *(even more for non industrial environments)*
SHIPMENT:	Ambient in storage containers
SAMPLE STABILITY:	Compound dependent (store @ -10 °C)
BLANKS:	1 to 3 per set
TECHNIQUE:	THERMAL DESORPTION, GAS CHROMATOGRAPHY, MASS SPECTROMETRY
ANALYTE:	See Table 1.1
DESORPTION:	Thermal desorption
INJECTION VOLUME:	Defined by desorption split flows (See Appendix)

TEMPERATURE	-DESORPTION:	300 °C for 10 min
	-DETECTOR (MS):	280 °C
	-COLUMN:	35 °C for 4 min; 8 °C/min to 150 °C, 15 °C/min to 300 °C
CARRIER GAS:		Helium
COLUMN:		30 meter DB-1, 0.25-mm ID, 1.0-µm film *(or equivalent)*
CALIBRATION:		Identification based on mass spectra interpretation and computerized library searches.
RANGE:		not applicable
ESTIMATED LOD:		100 ng per tube or less
PRECISION (S_r):		not applicable

- Personal sampling pump, 0.01 to 0.05 L/min, with flexible tubing.
- Shipping containers for thermal desorber sampling tubes.
- Instrumentation: thermal desorption system, focusing capability, desorption temperature appropriate to sorbents in tube (~300° C), and interfaced directly to a GC-MS system.
- Gas chromatograph with injector fitted with 1/4" column adapter, 1/4" Swagelok nuts and Teflon ferrules (or equivalent).
- Syringes: 1-µL, 10-µL (liquid); 100-µL, 500-µL (gas tight).
- Volumetric Flasks, 10-mL.
- Gas bulb, 2 L.

SPECIAL PRECAUTIONS: Some solvents are flammable and should be handled with caution in a fume hood. Precautions should be taken to avoid inhalation of the vapours from solvents as well. Skin contact should be avoided.

SAMPLING:

NOTE: Prior to field use, clean all thermal desorption tubes thoroughly by heating at or above the intended tube desorption temperature for 1-2 hours with carrier gas flowing at a rate of at least 50 mL/min. Always store tubes with long-term storage caps attached, or in containers that prevent contamination. Identify each tube uniquely with a permanent number on either the tube or tube container. Under no circumstances should tape or labels be applied directly to the thermal desorption tubes.

- Calibrate each personal sampling pump with a representative sampler in line.
- Remove the caps of the sampler immediately before sampling. Attach sampler to personal sampling pump with flexible tubing.
- NOTE: With a multi-bed sorbent tube, it is extremely important to sample in the correct direction, from least to maximum strength sorbent.
- For general screening, sample at 0.01 to 0.05 L/min for a maximum sample volume of 6 L. Replace caps immediately after sampling. Keep field blanks capped at all times. Tubes can act as diffusive samplers if left uncapped in a contaminated environment.
- Collect a "humidity test" sample to determine if the thermal adsorption tubes have a high water background.
- NOTE: At higher sample volumes, additional analyte and water (from humidity) may be collected on the sampling tube. At sufficiently high levels of analyte or water in the sample, the mass spectrometer may malfunction during analysis resulting in loss of data for a given sample.
- Collect a "control" sample. For indoor air samples this could be either an outside sample at the same location or an indoor sample taken in a non-complaint area.

- Ship in sample storage containers at ambient temperature. Store at -10° C.

SAMPLE PREPARATION:

- Allow samples to equilibrate to room temperature prior to analysis. Remove each sampler from its storage container.

- Analyze "humidity test" sampler first to determine if humidity was high during sampling (step 10).

- If high humidity, dry purge the tubes with purified helium at 50 to 100 mL/min for a maximum of 3 L at ambient temperature prior to analysis..

- Place the sampler into the thermal desorber. Desorb in reverse direction to sampling flow.

CALIBRATION AND QUALITY CONTROL:

- Tune the mass spectrometer according to manufacturer's directions to calibrate.

- Make at least one blank run prior to analyzing any field samples to ensure that the TD-GC-MS system produces a clean chromatographic background. Also make a blank run after analysis of heavily concentrated samples to prevent any carryover in the system. If carryover is observed, make additional blank runs until the contamination is flushed from the thermal desorber system.

- Maintain a log of thermal desorber tube use to record the number of times used and compounds found. If unexpected analytes are found in samples, the log can be checked to verify if the tube may have been exposed to these analytes during a previous sampling use.

- Run spiked samples along with the screening samples to confirm the compounds of interest. To prepare spiked samples, use the procedure outlined in the Appendix.

MEASUREMENT:

- See Appendix for conditions. MS scan range should cover the ions of interest, typically from 20 to 300 atomic mass units (amu). Mass spectra can either be identified by library searching or by manual interpretation (see Table 2). In all cases, library matches should also be checked for accurate identification and verified with standard spikes if necessary.

EVALUATION OF METHOD:

The method has been used for a number of field screening evaluations to detect volatile organic compounds. Estimate of the limit of detection for the method is based on the analysis of spiked samples for a number of different types of organic compounds. For the compounds studied, reliable mass spectra were collected at a level of 100 ng per compound or less. In situations where high levels of humidity may be present on the sample, some of the polar volatile compounds may not be efficiently collected on the internal trap of the thermal desorber. In these situations, purging of the samples with 3 L of helium at 100 mL/min removed the excess water and did not appreciably affect the recovery of the analytes on the sample.

Table 2. Common volatile organic compounds with mass spectral data.

Compound/ Synonyms	CAS# RTECS	Empirical Formula	MW[a]	BP[b] (°C)	VP[c] @ 25°C mm Hg	kPa	Characteristic Ions, m/z
Aromatic Hydrocarbons							
Benzene /benzol	71-43-2 CY1400000	C_6H_6	78.11	80.1	95.2	12.7	78*
Xylene /dimethyl benzene	1330-20-7 ZE2100000	C_8H_{10}	106.7				91, 106*, 105
o-xylene				144.4	6.7	0.9	
m-xylene				139.1	8.4	1.1	
p-xylene				138.4	8.8	1.2	
Toluene /toluol	108-88-3 XS5250000	C_7H_8	92.14	110.6	28.4	3.8	91, 92*

Aliphatic Hydrocarbons							
n-Pentane	109-66-0 RZ9450000	C_5H_{12}	72.15	36.1	512.5	68.3	43, 72*, 57
n-Hexane /hexyl-hydride	110-54-3 MN9275000	C_6H_{14}	86.18	68.7	151.3	20.2	57, 43, 86*, 41
n-Heptane	142-82-5 MI7700000	C_7H_{16}	100.21	98.4	45.8	6.1	43, 71, 57, 100*,41
n-Octane	111-65-9 RG8400000	C_8H_{18}	114.23	125.7	14.0	1.9	43, 85, 114*, 57
n-Decane /decyl hydride	124-18-5 HD6500000	$C_{10}H_{22}$	142.29	174	1.4	0.2	43, 57, 71, 41, 142*
Ketones							
Acetone /2-propanone	67-64-1 AL3150000	C_3H_6O	58.08	56	266	35.5	43, 58*
2-Butanone /methyl ethyl ketone	78-93-3 EL6475000	C_4H_8O	72.11	79.6	100	13	43, 72*
Methyl isobutyl ketone /MIBK, hexone	108-10-1 SA9275000	$C_6H_{12}O$	100.16	117	15	2	43, 100*, 58
Cyclohexanone /cyclohexyl ketone	108-94-1 GW1050000	$C_6H_{10}O$	98.15	155	2	0.3	55, 42, 98*, 69
Alcohols							
Methanol /methyl alcohol	67-56-1 PC1400000	CH_3OH	32.04	64.5	115	15.3	31, 29, 32*
Ethanol /ethyl alcohol	64-17-5 KQ6300000	C_2H_50H	46.07	78.5	42	5.6	31, 45, 46*
Isopropanol /1-methyl ethanol	67-63-0 NT8050000	C_3H_7OH	60.09	82.5	33	4.4	45, 59, 43
Butanol /butyl alcohol	71-36-3 E01400000	C_4H_9OH	74.12	117	4.2	0.56	56, 31, 41, 43
Glycol Ethers							
Butyl cellosolve /2-butoxyethanol	111-76-2 KJ8575000	$C_6H_{14}O_2$	118.17	171	0.8	0.11	57, 41, 45, 75, 87
Diethylene glycol ethyl ether /Carbitol	111-90-0 KK8750000	$C_6H_{14}O_3$	134.17	202	0.08	0.01	45, 59, 72, 73, 75, 104
Phenolics							
Phenol /hydroxybenzene	108-95-2 SJ3325000	C_6H_5OH	94.11	182	47	0.35	94*, 65, 66, 39
Cresol	1319-77-3 G05950000	C_7H_7OH	108.14				108*, 107, 77, 79
2-methylphenol	95-48-7			190.9	1.9	0.25	
3-methylphenol	108-39-4			202.2	1.0	0.15	
4-methylphenol	106-44-5			201.9	0.8	0.11	

Chlorinated Hydrocarbons							
Methylene chloride /dichloromethane	75-09-2 PA8050000	CH_2Cl_2	84.94	40	349	47	86*, 84, 49, 51
1,1,1-Trichloroethane /methyl chloroform	71-55-6 KJ2975000	CCl_3CH_3	133.42	75	100	13.5	97, 99, 117, 119
Perchloroethylene /hexachloroethane	127-18-4 KX3850000	CCl_3CCl_3	236.74	187 (subl)	0.2	<0.1	164*, 166, 168, 129, 131, 133, 94, 96
o-,p-Dichlorobenzenes		$C_6H_4Cl_2$	147.0				146*, 148, 111, 113, 75
/1,2-dichlorobenzene	95-50-1 CZ4500000			172-9	1.2	0.2	
/1,4-dichlorobenzene	106-46-7 CZ4550000			173.7	1.7	0.2	
1,1,2-Trichloro-1,2,2-trifluoroethane /Freon 113	76-13-1 KJ4000000	CCl_2FCClF_2	187.38	47.6	384	38	101, 103, 151, 153, 85, 87
Terpenes							
d-Limonene	5989-27-5 OS8100000	$C_{10}H_{16}$	136.23	176	1.2		68, 67, 93, 121, 136*
Turpentine (Pinenes)	8006-64-2	$C_{10}H_{16}$	136.23	156 to 170	4 @ 20°		93, 121, 136*, 91
α-pinene	80-56-8			156			
β-pinene	127-91-3			165			
Aldehydes							
Hexanal /caproaldehyde	66-25-1 MN7175000	$C_6H_{12}O$	100.16	131	10	1.3	44, 56, 72, 82, 41
Benzaldehyde /benzoic aldehyde	100-52-7 CU4375000	$C_7H_{12}O$	106.12	179	1.0	0.1	
Nonanal /pelargonic aldehyde	124-19-6 RA5700000	$C_9H_{18}O$	142.24	93	23	3	43, 44, 57, 98, 114
Acetates							
Ethyl acetate /acetic ether	141-78-6 AH5425000	$C_4H_8O_2$	88.1	77	73	9.7	43, 88*, 61, 70, 73, 45
Butyl acetate /acetic acid butyl ester	123-86-4 AF7350000	$C_6H_{12}O_2$	116.16	126	10	1.3	43, 56, 73, 61
Amyl acetate /banana oil	628-63-7 AJ1925000	$C_7H_{14}O_2$	130.18	149	4	0.5	43, 70, 55, 61
Other							
Octamethylcyclotetra-siloxane	556-67-2 GZ4397000	$C_8H_{24}O_4Si_4$	296.62	175			281, 282, 283

[a] Molecular Weight
[b] Boiling Point
[c] Vapour Pressure
*Indicates molecular ion

APPENDIX

Multi-bed sorbent tubes: Other sorbent combinations and instrumentation/conditions shown to be equivalent may be substituted for those listed. In particular, if the compounds of interest are known, specific sorbents and conditions can be chosen that work best for that particular compound(s). The tubes that have been used in NIOSH studies with the Perkin Elmer ATD system are ¼ " stainless steel tubes, and are shown in the figure in the next page:

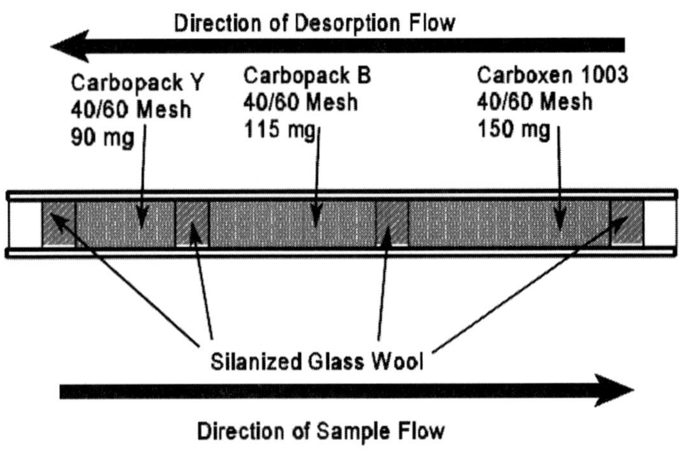

Figure 1. Carbopack™ and Carboxen™ adsorbents are available from Supelco, Inc.

Preparation of spiked samples: Spiked tubes can be prepared from either liquid or gas bulb standards.

***Liquid standards*:** Stock solutions are prepared by adding known amounts of analytes to 10-mL volumetric flasks containing high purity solvent (carbon disulfide, methanol, toluene). Solvents are chosen based on solubility for the analytes of interest and ability to be separated from the analytes when chromatographed. Highly volatile compounds should be dissolved in a less volatile solvent. For most compounds, carbon disulfide is a good general purpose solvent, although this will interfere with early eluting compounds.

Gas bulb standards: Inject known amounts of organic analytes of interest into a gas bulb of known volume filled with clean air. Prior to closing the bulb, a magnetic stirrer and several glass beads are placed in the bulb to assist in agitation after introduction of the analytes. After injection of all of the analytes of interest into the bulb, warm the bulb to 50 °C and place it on a magnetic stirring plate and stir for several minutes to ensure complete vaporization of the analytes. After the bulb has been stirred and cooled to room temperature, remove aliquots from the bulb with a gas syringe and inject into a sample tube as described below.

Tube spiking Fit a GC injector with a ¼" column adapter. Maintain the injector at 120°C to assist in vaporization of the injected sample. Attach cleaned thermal desorption tubes to injector with ¼" Swagelok nuts and Teflon ferrules, and adjust helium flow though the injector to 50 mL/min. Attach the sampling tube so that flow direction is the same as for sampling. Take an aliquot of standard solution (gas standards 100 to 500 μL; liquid standards, 0.1 to 2 μL) and inject into the GC injector. Allow to equilibrate for 10 minutes. Remove tube and analyze by thermal desorption using the same conditions as for field samples.

Instrumentation: Actual media, instrumentation, and conditions used for general screening of unknown environments are as follows: Perkin-Elmer ATD 400 (automated thermal desorption system) interfaced directly to a Hewlett-Packard 5980 gas chromatograph/HP5970 mass selective detector and data system.

ATD conditions:

Tube desorption temperature: 300°C

Tube desorption time: 10 min.

Valve/transfer line temperatures: 150°C

Focusing trap: Carbopack B/Carboxen 1000, 60/80 mesh, held at 27°C during tube desorption

Focusing trap desorption temperature: 300°C

Desorption flow: 50-60 mL/min.

Inlet split: off

Outlet split: 20 mL/min.

Helium: 10 PSI

GC conditions:

DB-1 fused silica capillary column, 30 meter, 1-µm film thickness, 0.25-mm I.D. Temperature program: Initial 35°C for 4 min, ramp to 100°C at 8 /min, then ramp to 300°C at 15 /min, hold 1-5 min.

Run time: 27 min.

MSD conditions:

Transfer line: 280°C

Scan 20-300 amus, EI mode

EMV: set at tuning value

Solvent delay: 0 min for field samples; if a solvent-spiked tube is analyzed, a solvent delay may be necessary to prevent MS shutdown caused by excessive pressure.

Alternative methods may be obtained adapting, for instance, NIOSH Method 1500 (Hydrocarbons b.p. 36 – 216°C) or NIOSH Method 1501 (Hydrocarbons, aromatic).

Main differences – not fundamentals – may consist in: sorbent choice (charcoal tube), desorption technique (elution with CS_2), sampling time (longer than in workplaces) and other. More important is the possibility of using diffusive (passive) samplers.

5. PASSIVE SAMPLING

The original purpose of the development on passive sampling (based on Fick's laws) is to provide technology at low cost, enabling air quality surveys to be routinely executed at multiple locations within urban and rural areas, industrial sites and forests. This requires the examination of the performance characteristics of the diffusive sampler over long sampling periods, in comparison with established methods, and in practical applications of urban monitoring (Brown et al., 1994; Brown et al., 1999). Several authors proposed the use of different models of diffusive (passive) devices (Brown, 1993; Harper, 2000; Bertoni et al., 2001; Brown, 2002; Kot-Wasik et al., 2007) conceived for the determination of long-term averaged concentration of some airborne volatile and semi-volatile organic compounds, relevant on the human health protection (VOCs, polycyclic aromatic hydrocarbons and nicotine). The diffusive sampling technique is known to be the cheapest and easiest way to perform extensive sampling campaigns, both in temporal and spatial terms. Moreover, this is the only collection technique allowing the true separation of the vapour phase species from the particle bound fraction (Bertoni et al., 2004; Namiesik et al., 2005).

A lot of different models of passive sampler have been proposed by many researchers; in this chapter we can show only a "classic" sampler (Analyst, developed by the Italian National Research Council, IIA-CNR, Fig. 2 and 3) and a more recent sampler, based on radial diffusion principle (Ring, IIA-CNR, Fig. 4).

All the diffusive samplers have to respect the requirements and test methods according to EN 13528-1,EN 13528-2, EN 13528-3, EN 14412 and EN-ISO 16017-2.

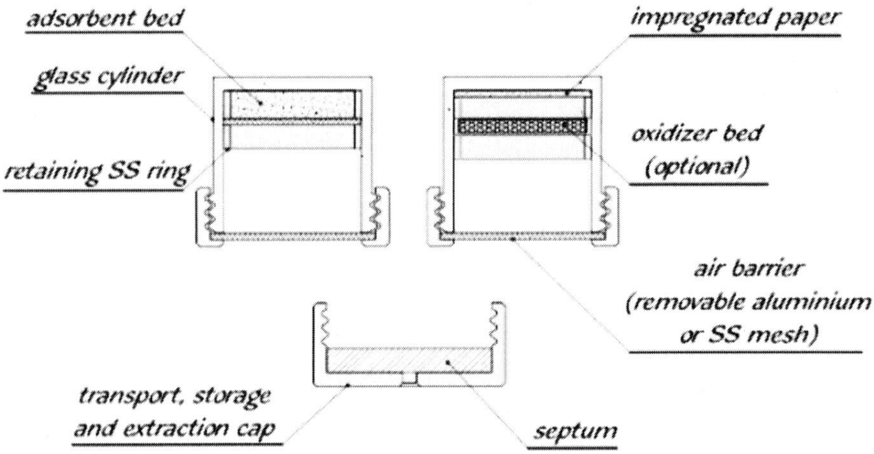

Figure 2. Analyst sampler scheme

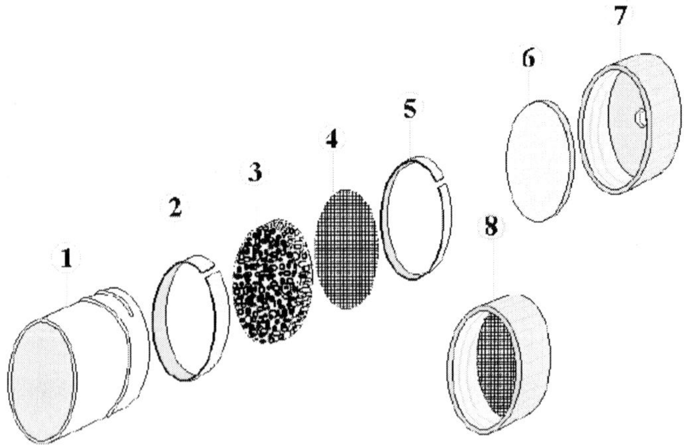

Figure 3. Scheme of the Analyst : 1: glass cylinder (i.d. = 20 mm); 2 and 5: retaining S.S. rings; 3 adsorbent bed; 4: viewing S.S. ring; 6: Teflon seal; 7: cap; 8: aluminium diffusion cap.

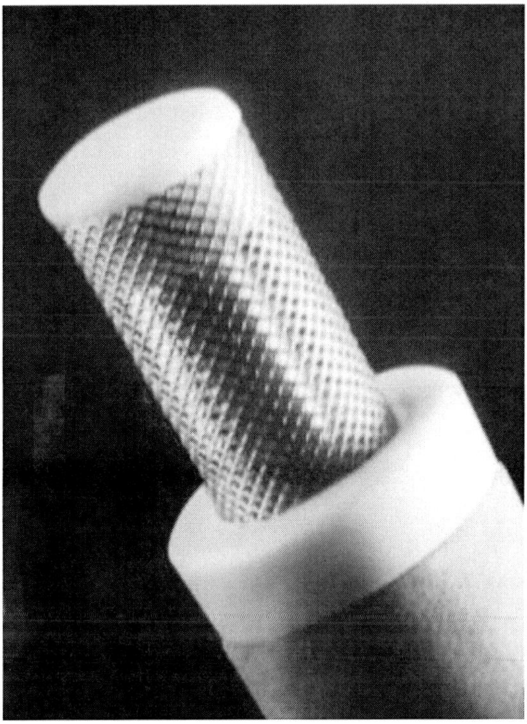

Figure 4. Diffusive sampler "Ring" (IIA CNR).

6. DATA EVALUATION AND EXPOSURE RISK ASSESSMENT

Studies have found that home levels of several organics average 2 to 5 times higher indoors than outdoors. During and for several hours immediately after certain activities, such as paint stripping, levels may be 1,000 folds background outdoor levels. Despite direct reading instruments can be used (like the IR used in indoor monitoring shown in Figure 5), as above mentioned, currently it is better to try to obtain very low LODs and LOQs by employing sampling methods as described.

Sometimes they are used Threshold Limit Values (TLVs) that are the guideline values set by the American Conference of Governmental

Industrial Hygienists (ACGIH) to minimize workers exposure to hazardous concentrations as much as possible. The TLVs are published yearly for more than 700 chemical substances and physical agents (ACGIH, 2010). This use is incorrect and unauthorized by ACGIH, because the TLVs respect is not a warranty of no health effects, that is on the contrary a fundamental requirement in nonindustrial environment.

No standards have been set for VOCs in non-industrial settings. OSHA regulates formaldehyde, a specific VOC, as a carcinogen. OSHA has adopted a Permissible Exposure Level (PEL) of 0.75 ppm, and an action level of 0.5 ppm. Based upon current information, it is advisable to mitigate formaldehyde that is present at levels higher than 0.1 ppm. Recently the World Health Organization (WHO, 2010) provided health-based guidelines for 55 airborne inorganic and organic compounds for carcinogenic and non-carcinogenic health endpoints. The non-carcinogenic endpoints include development toxicity, reproduction toxicity, respiratory toxicity, neurotoxicy, hepatoxicity, hematoxicity, eye/nose/throat irritation, and odor annoyance. The lowest concentration at which effects are observed in humans, animals, and plants was used as a staring point for the non-carcinogenic endpoints. Uncertainty factors determined through scientific judgment in consensus and averaging time were also taken into account in determining the health endpoint for non-carcinogenic compounds. The classification by the international Agency for Research on Cancer (IARC) was used to determine a chemical as a carcinogen. The endpoint of carcinogen was determined by linear extrapolation from the high dose level, which is characteristic of animal experiments or occupation exposure with cancer responses.

This objective requires measurements for checking whether specified limit or guideline values are being exceeded. Examples of limit and guideline values (except workplace atmospheres and ambient air) for indoor environments are given in Table 3 and 4, established by the ad hoc committee of IRK/AOLG of Germany for organic chemicals.

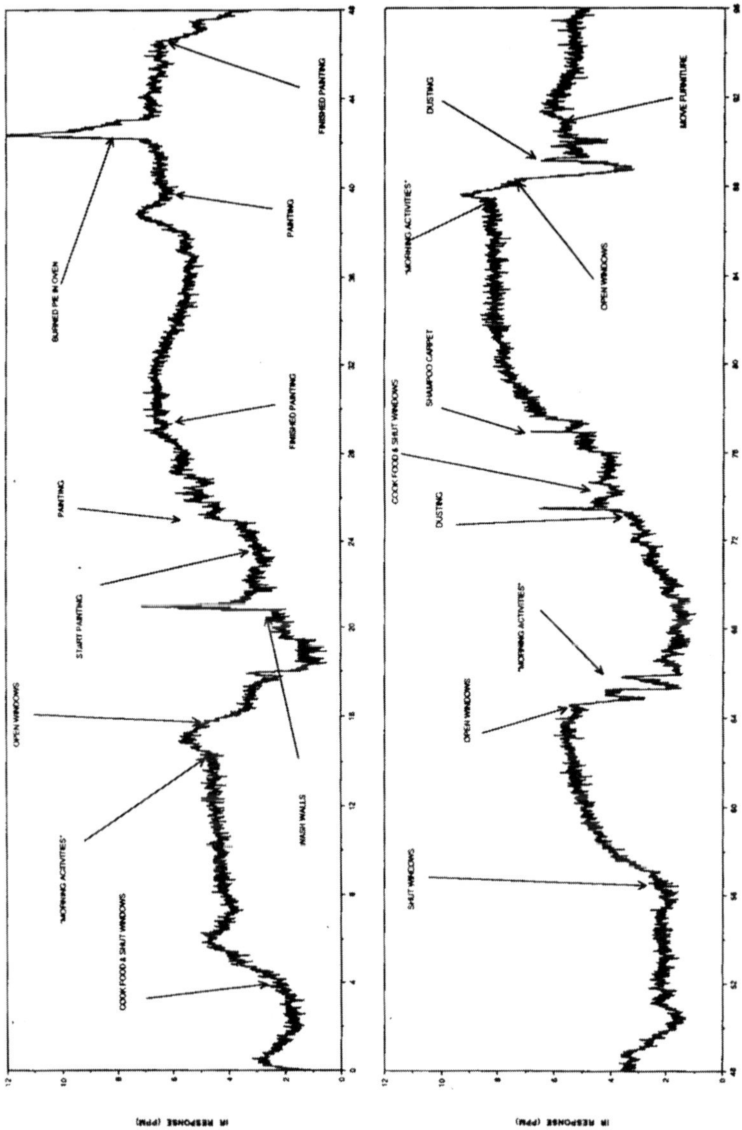

Figure 5. Variation in indoor VOC levels in house house as detected by an IR instrument. VOC levels are expressed in ppm of equivalent 33% propane/67% butane. Human activities at different times are shown (Clobes et al., 1992).

Table 3. Examples of limit values.

Pollutant	Country	Limit value	Pollutant
Tetrachloroethene	Germany	0.1 mg/m³	7 days

Table 4. Examples of guideline values.

Pollutant	Country/Organisation	Guideline Value	Averaging Time
Formaldehyde	Germany	0.12 mg/m³	not specified
Nitrogen dioxide	Germany	60 µg/m³ [a]	7 days
Toluene	Germany	0.3 mg/m³ [a]	not specified
Styrene	Germany	0.03 mg/m³ [a]	7 days
Dichloromethane	Germany	0.2 mg/m³ [a]	24 h
TVOC	Germany	0.2 – 0.3 mg/m³	not specified

[a] Guideline value (RW I-value, "Richtwert I"= guideline value I) aimed at hygienic prevention.

REFERENCES

1 American Conference of Governmental Industrial Hygienists (ACGIH) 2010 Threshold Limit Values and Biological Exposure Indices, Cincinnati OH- USA.

2 G. Bertoni, R. Tappa, I. Allegrini, 2001 "The internal consistency of the 'Analyst' Diffusive sampler- a long-term field test"- Chromatographia 54, November (9 10) 653- 657

3 G. Bertoni, R. Tappa, A. Cecinato, 2001 "Environmental Monitoring of semi-volatile polyciclic aromatic hydrocarbons by means of diffusive sampling devices and GC-MS analysis" Chromatographia, 53, Suppl., S312 - S316.

4 G. Bertoni, C. Ciuchini, R. Tappa, 2004 "Long-term diffusive samplers for the indoor air quality evaluation",- Ann. Chim. 94 (Rome), 637644 .

5 R. H. Brown, 2002 Monitoring volatile organic compounds in air- the development of ISO standards and a critical appraisal of the methods- J Environ Mon 4, 112N-118N

6 R. H. Brown, M. D. Wright, N. T. Plant, 1999 The Use of Diffusive Sampling for Monitoring of Benzene, Toluene and Xylene in Ambient Air (IUPAC Technical Report) Pure Appl. Chem., 71 10 19932008 .

7 R. H. Brown, 1993 The use of diffusive samplers for monitoring of ambient air- Pure & Appl. Chem. 65 18591874

8 R. H. Brown, R. P. Harvey, C. J. Purnall, K. J. Saunders, 1984 A Diffusive Sampler Evaluation Protocol- Am. Ind. Hyg. Assoc. J. 45, 67 EOF75 EOF .

9 A. L. Clobes, G. P. Ananth, A. L. Hood, J. A. Schroeder, K. A. Lee, 1992 Human Activities as Source of Volatile Organic Compounds in Residential Environments; in "Sources of Indoor Air Contaminants", Tucker W.G., Leaderer B.P., Molhave L. and Cain W.S. editors. Annals of the New York Academy of Sciences, 641 79-86.

10 M. Harper, trapping. Review-Sorbent, volatile. of, compounds. organic, air. from-J, Chrom, 885 885 2000 129151

11 A. Kot-Wasik, B. Zabiegala, M. Urbanovicz, E. Dominiak, A. Wasik, J. Namiesnik, 2007 Advances in passive sampling in environmental studies- Anal Chim Acta 602 141164

12 M. Maroni, R. Axelrad, A. Bacaloni, 1995 "NATO's Efforts to set IAQ Standards" Am. Ind. Hygiene Ass. J., 56 499508

13 J. Namiesnik, Z. Bozena, A. Kot-Wasik, M.. Partyka, A. Wasik, 2005 Passive sampling and/or extraction techniques in environmental analysis: a review- Anal Bioanal Chem 381 279301

14 EN 13528- 1 (Sept. 2002) Ambient air quality: Diffusive samplers for the determination of concentrations of gases and vapours-Requirements and test methods. Part 1: General requirements

15 EN 13528- 2 (Sept. 2002) Ambient air quality: Diffusive samplers for the determination of concentrations of gases and vapours- Requirements and test methods. Part 3: Specific requirements and test methods

16 EN 13528- 3 (Dec. 2003) Ambient air quality: Diffusive samplers for the determination of concentrations of gases and vapours- Requirements and test methods. Part 3: Guide to selection, use and maintenance

17 EN 14412 (Sept. 2004) Indoor air quality- Diffusive samplers for the determination of concentrations of gases and vapours- Guide for selection, use and maintenance

18 EN ISO 16017-2 (May 2003) Indoor, ambient and workplace air- Sampling and analysis of volatile organic compounds by sorbent tube/thermal desorption/capillary gas chromatography. Part 2: Diffusive sampling

19 NIOSH Manual of Analytical Methods, 1996 Method 2549 Volatile Organic Compounds (screening) 4th rev., 1-ed . Cincinnati, OH- USA.

20 NIOSH Manual of Analytical Methods, 2003 Method 1500 Hydrocarbons b.36216 C, 3-ed . Cincinnati, OH- USA.

21 NIOSH Manual of Analytical Methods, 2003 Method 1501 Hydrocarbons, aromatic, 3-ed . Cincinnati, OH- USA.

22 WHO guidelines for indoor air quality: selected pollutants 2010 978-9-28900-213-4

23 http://www.epa.gov/iaq/atozindex.html

24 http://www.umweltbundesamt.de/uba-info-daten-e/daten-e/irk.htm#4

25 http://www.umweltbundesamt.de/uba-info-daten/daten/irk.htm

26 http://www.osha.gov/dts/osta/otm/otm_iii/otm_iii_2.html

CHAPTER 6

Effectiveness of an Indoor Air Pollution (IAP) Intervention on Reducing IAP and Improving Women's Health Status in Rural Areas of Gansu Province, China

Yibin Cheng[1*], Jiaqi Kang[1*], Fan Liu[1], Bryan A. Bassig[2], Brian Leaderer[2], Gongli He[1], Theodore R. Holford[2], Ning Tang[1], Jian Wang[3], Jian He[4], Yanchang Liu[1], Yingchun Liu[1], Jiang Liu[1], Xun Chen[1], Heng Gu[1], Xiao Ma[5], Tongzhang Zheng[2], Yinlong Jin[1]

[1]Institute for Environmental Health and Related Product Safety, Chinese Center for Disease Control and Prevention, Beijing, China
[2]Yale School of Public Health, New Haven, USA
[3]Chinese Center for Disease Control and Prevention, Beijing, China
[4]Gansu Provincial Center for Disease Control and Prevention, Lanzhou, China
[5]West China School of Public Health, Sichuan University, Chengdu, China

ABSTRACT

Given the deleterious health effects associated with indoor air pollution (IAP), this study was conducted to evaluate an IAP intervention in rural areas in Gansu, one of the poorest provinces of China. We selected 371 rural households to take part in intervention measures including stove improvement and health education. Eight of 371 households were selected to conduct IAP sampling. Four hundred and thirteen women in these households completed a questionnaire and 49 women took part in lung function tests. After the intervention, PM_4 levels reduced from 455 μg/m^3 to 200 μg/m^3 and CO reduced from 3.40 ppm to 2.90 ppm in indoor air. The percentage of predicted value of FEV1 and FVC improved to some degree after the intervention, but all the parameters of lung function assessment did not show a significant change. Prevalence rates of several symptoms associated with IAP significantly declined in the study population, compared with baseline levels. Intervention measures combining stove improvement with health education were effective in reducing IAP levels. Women's health status, including eye and respiratory symptoms, also showed improvement. However, the effect on lung function was not apparent and warranted additional follow-up. Similarly, evaluation of the long term effects of the IAP intervention will require future studies.

KEYWORDS

Indoor Air Pollution, Biomass, Lung Function, Respiratory System

1. INTRODUCTION

About 41% of the world's population use solid fuel (such as coal and biomass) for domestic cooking and heating, and a large proportion of this exposed population lives in less developed countries [1] . Globally, reliance on solid fuels has emerged as one of the ten most important threats to public health [2] . In 2010, household air pollution from solid fuels was among the top 3 risk factors contributing to the global burden of disease and was responsible for nearly 3.5 million deaths across the world [3]. These deaths occur predominantly in women and children, as women are normally responsible for food preparation and cooking, and infants and young children are usually with their mothers near the cooking area and therefore are more likely to be exposed. In addition to deaths, various health issues, such as respiratory diseases, lung function reduction, and eye infections, have been associated with IAP exposure as recently reviewed or reported by others [4] - [6] . For example, exposure to high levels of SO_2 in the kitchen has been associated with a significantly higher prevalence of COPD among nonsmoking women [7] . Furthermore, women living in households that used biomass had a significantly higher prevalence of asthma than those in households using cleaner fuels, and the risk for asthma appeared to be higher for women than for men [8] .

More than half of the Chinese population lives in rural areas where more than 80% of the households use solid fuels for cooking [9]. WHO estimates of the disease burden associated with this use in China had suggested that solid fuel use was responsible for about 342,450 COPD deaths and 17,720 lung cancer deaths for those aged ≥ 30, and was also responsible for 3,204,900 total DALYs in China in 2002, which was equivalent to 1.6% of the national disease burden during that year [2] . A further 28% of global deaths caused by indoor smoke from solid fuels occur in China [10] .

In 2002, the World Bank, in conjunction with the Institute of Environmental Health and Related Product Safety of the Chinese Center for Disease Control and Prevention (IEHS, China CDC), initiated an IAP intervention product called the Sustainable and Efficient Energy Use to Alleviate Indoor Air Pollution in Poor Rural Areas of China. The purpose of the project was to test the viability of both technological and behavioral interventions to mitigate IAP and improve human health, and the project was carried out in rural areas of 4 Chinese remote provinces including Gansu, Guizhou, Shaanxi, and Inner Mongolia. The IAP intervention involved stove improvement in each selected area and health education (behavior intervention) of the local residents with the goal of changing their traditional cooking methods, heating practices, and other lifestyles that might contribute to IAP [11] [12] .

Some previous epidemiologic evidence has indicated the beneficial effects of stove improvement as far as associated reduced risks for lung cancer, pneumonia, and COPD in other rural areas in China following implementation, which suggests that such an intervention may be viable [13] - [15] . Here, we report the results assessing the effectiveness of the IAP intervention on reducing IAP and improving the health status in women in Hui County of Gansu Province. We report the effects of the intervention on multiple indices related to IAP mitigation, including changes in various symptoms of eye and respiratory diseases, measures of lung function, and assessment of PM_4 and CO levels in a representative sample of the households both before and after the intervention period.

2. MATERIALS AND METHODS

2.1. Study Population

Gansu, located in northwest China, is one of China's least developed provinces. The annual income per capita for rural residents in the province was estimated to be around US $193 in 2002 according to the National Bureau of Statistics of China. Wood and crop residues are the main energy sources for domestic cooking and heating in this area. The regular heating season in the region is between November and April of the following year. Hui County was selected for the stove intervention study because households in this county commonly use traditional stoves with biomass for heating and cooking, which causes direct release of smoke into the indoor living environment. While some households had chimneys, the chimneys were installed too low (for example, lower than eave) and consequently the smoke can easily reenter into rooms and result in IAP. Moreover, very few families in the area installed ventilators in their kitchen.

A total of 413 women from 371 households were included in the stove intervention study. Households that met the following criteria were recruited for this study: 1) used biomass for heating and cooking before the intervention with traditional stoves, e.g. open fire, unvented stoves, or stoves without a smoke door; 2) included woman aged 18 and over and children under age 15; 3) had residents who lived in Hui County continuously for more than one year at the time that the stove intervention began.

2.2. Stove Improvement

Stove types used in homes in this region before the intervention consisted of either open pits (Figure 1(a)) or hand stoves (Figure 1(b)), or traditional stoves with either no chimney in the home (Figure 1(c)) or a chimney that was installed too low to effectively mitigate IAP (Figure 1(d)). The improved stoves with a chimney that were used for the intervention (Figure 2(a) and Figure 2(b)) were specifically designed for the local residents and were pilot tested for effectiveness before the stove intervention, as described elsewhere [16] . The local technicians were trained and certified by the Chinese NIEHS and were responsible for the installation of the stoves and for training each of the residents on how to use the stoves. Because some of the traditional stoves were used by the subjects for heating, the improved stove was connected to a heating bed (called Kang as shown inFigure 2(c)) and thus could be used for heating purposes as well. Results from the pilot tests showed that emission of PM_4 and CO reduced by 98% and 88%, respectively, in controlled conditions [16] . Between August and November in 2004, all 371 participating families changed to a certified stove with a chimney for cooking and heating. The old stoves inside these homes were completely demolished and replaced by the improved stoves. All homes in this region use agricultural residue and wood as their primary source of fuel for cooking and heating, and these fuel types did not change over the course of this study.

Figure 1. (a)-(d) Stoves used by the households in the study area before the intervention. Stoves used in this region before the intervention included either: open pits (a) or hand stoves (b), or traditional stoves with either no chimney in the home (c) or a chimney that was installed too low to effectively mitigate IAP (d).

Figure 2. Improved stoves used by the households in the study area after the intervention. The improved stoves with a chimney that were used for the intervention (a) and (b) were specifically designed for the local residents. The improved stove was connected to a heating bed (called Kang as shown in (c).

2.3. Health Education (Behavior Intervention)

To encourage residents to actively participate in the stove improvement activities and correctly use the newly installed stoves, and to change lifestyle habits that may contribute to IAP or increasing exposure to IAP, scientists from IEHS, China CDC, and the Huaxi School of Public Health, Sichuan University, trained and worked with the local health professionals, village physicians, local stove technicians, and school teachers to educate the local residents during 2003 and 2005. This component included providing knowledge on the sources and impacts of IAP and information on correctly using and maintaining the stoves. Pamphlets describing methods to prevent IAP were distributed during household visits, health education classes, and other related activities. Three specific factors were evaluated to measure the impact of the behavior intervention, including use of a fan in the kitchen, opening of windows during the winter, and reduction in use of portable open fires to warm the pot tea.

2.4. Assessment of Health and Intervention Effects

Following informed consent, information on respiratory symptoms and occurrence of eye irritation/infections were collected through in-person interviews conducted by 30 trained interviewers. In both pre-intervention baseline interviews (during April, 2003) and post-intervention interviews (during April, 2005), participating women were asked to report their respiratory and eye symptoms during the 3 months before the interviews. The post- intervention health assessments, including lung function testing, were conducted between 5 - 8 months after the improved stoves were installed during August-November 2004. Information on the type and amount of energy use, stove type, cooking habits, ventilation, heating, and other lifestyle and demographic data was also collected during the in-person interviews. A total of 49 women participated in both pre- and post-intervention lung function testing. These 49 women were similar to the overall study population with respect to demographic characteristics and cooking practices, as well as characteristics associated with the home, such as house area and presence of a short chimney. The pre-intervention spirometry was performed in April 2003 and post-intervention spirometry was conducted in April 2005 by technicians from IEHS, China CDC using a Multi-Functional Spirometer HI-801 (Chest M.I., INC, Tokyo, Japan). These technicians were trained before the field component of the study in accordance with the standardized steps of the ATS/ERS Task Force: Standardization of Lung Function Testing. The collected measurements on lung function parameters included vital capacity (VC), forced expiratory volume in 1 second (FEV1), forced vital capacity (FVC), and forced expiratory flow 25% - 75% (FEF 25% - 75%). All lung function tests were conducted by trained personnel in the morning during the study period. For each of the lung function parameters, three measurements were taken and the best result of the 3 measurements was recorded and reported here.

2.5. Assessment of Indoor Air Pollutants

Because of the cost of monitoring and the willingness of the households, eight households were selected for IAP monitoring. These 8 families were selected to represent the geographic location, stove type, and type of biomass used in the region. PM_4 and CO in the kitchen and bedroom were measured before (March 2003) and between 5-8 months after the stove intervention (March 2005). Indoor air samples were collected consecutively for 24 hours in each of the 8 households. Respirable particles were measured according to the National Institute for Occupational Safety and Health of the United States (NIOSH, USA) protocol 0600, designed to capture particles with a median aerodynamic diameter of 4 ím (PM4). Samples were collected using a 10-mm nylon cyclone equipped with a 37-mm diameter poly vinyl chloride (PVC) filter (pore size 5ím supplied by SKC Inc., USA) at a flow rate of 2.5 l/min. Air was drawn through the cyclone preselectors using battery-operated constant flow pumps (model PCXR8 supplied by SKC Inc., USA). All pumps were calibrated prior to and after each sampling day using a field minimeter, itself calibrated by a soap bubble meter in the laboratory. Pumps were also calibrated in the laboratory after each field exercise using the same minimeter. To maintain battery power throughout the sampling period, pumps were programmed to cover the 24-h interval through intermittent sampling (1 min out of every 4 - 6 min). One field blank was taken on each sampling day.

Gravimetric analyses were conducted at the laboratory of the National Institute for Environmental Health and Related Products Safety, China CDC using an analytic microbalance (1/100,000, Sartorius 2004 MP, Germany) calibrated against standards provided by the Bureau of National Technological Control. All filters (field blanks and samples) were conditioned for 24 hours before weighing. The weighing facility temperature was set at 20 - 25 degrees C and humidity at 50% ± 5%. Respirable dust concentrations were calculated by dividing the blank-corrected increase in filter mass by the total air volume sampled [17] . Carbon monoxide (CO) was measured using long term diffusion tubes (manufactured by GASTEC, USA), with detection ranges of 10 - 200 or 50 - 1000 ppm [17] .

2.6. Statistical Analysis

The 24 hour personal exposure to PM_4 and CO in kitchen and bedroom air was compared for the pre- and post- IAP intervention periods using the Wilcoxon signed rank test. Changes in lifestyles before and after the intervention were compared using McNemar's test. Lung function was assessed by calculating means and standard deviations based on observed values and the percentage of predicted values of each parameter. Predicted values were the average observed values in the population for any person of similar age, sex, and body composition, calculated in terms of predict equation for the Chinese population [18] . The percentages of predicted values were the ratios of observed values divided by the predicted value so that age, sex, and body composition would be adjusted by comparing the relative values. In addition to the possible change of body composition, we used linear mixed models to adjust for other factors including cooking years up to the date of survey, use of an open fire, not using fans in the kitchen, location of the kitchen in the home, and whether or not windows were opened when cooking. For the total study population (413 women), the effectiveness of the IAP intervention on reducing the respiratory and eye irritation symptoms was evaluated by logistic regression, adjusting for age, cooking years up to the date of the survey, using open fire, location of the kitchen, and never opening windows when cooking. For 49 spirometry participants, McNemar's test was used for the analysis of symptoms.

3. RESULTS

3.1. Characteristics of Study Population

Characteristics of the study population based on data from the baseline survey are shown in Table 1. Over 80% of the subjects were between 18 - 39 years old and 99.8% of subjects were Han, the main ethnicity of China. Over 50% of the participants lived in households with 5 or more family members, and biomass was the main living fuel in all the selected households. 98.1% of women were responsible for cooking and 63.7% of subjects reported an average daily cooking time of 2 to 4 hours. None of the subjects were smokers, but 70% of them reported exposure to second-hand smoke. Very few women reported a change in their passive smoking exposure status after the intervention, compared to before because the interviews took place after such a short period following the stove improvement. The prevalence rates for all of the considered chronic diseases were low and were less than 5% in the study population.

3.2. Changes of Stove Features and Behaviors

The changes in stove features and behaviors relating to stove use associated with the intervention are shown in Table 2. Before the intervention, the percentages of stoves without chimneys, short chimneys (i.e., smoke exit lower than eaves), and stoves without smoke doors were 6.1%, 71.7% and 87.7%, respectively. These traditional designs of the stoves were considered primary contributors to IAP in households that used biomass. Following the intervention, the percentages of stoves having each of these characteristics reduced to 0% (Table 2).

Table 1. Characteristics of the study population (%).

Variables	n	Percentage %
Age (baseline)		
18 - 29	109	26.4
30 - 39	230	55.7
40 - 49	32	7.8
50 - 59	31	7.5
60~	11	2.7
Ethnicity		
Han	412	99.8
Other	1	0.2
Family members		
3	45	10.9
4	135	32.7
5	112	27.1
6	84	20.3
>6	36	8.7
Missing	1	0.2
Biomass use	413	100.0
Cooking	405	98.1
Everyday cooking time (hrs)		
<2	105	25.4
2~	263	63.7
4~	39	9.4
6~	6	1.5
Smoking	0	0
Passive smoking	289	70.0
History of diagnosed chronic diseases (%)		
Rhinitis	3	0.7
Faucitis	2	0.5
Tuberculosis	2	0.5
Asthma	1	0.2
Emphysema	0	0
Chronic bronchitis	9	2.2
Hypertension	12	2.9
Heart disease	14	4.1
Allergy	4	1.0

Table 2. Changes of stove features and behaviors before and after intervention (%).

Stove features and behaviors	Before (n = 413)	After (n = 413)	p-value[b]
Stoves without chimneys	6.1	0	—
Short chimney (exit of chimney lower than eave)[a]	71.7	0	—
Stoves without smoke doors	87.7	0	—
Portable open fire used for heating tea	35.6	33.7	0.01
Fans installed in kitchen	1.0	5.6	<0.01
Never open window in winter	16.0	11.1	0.048

[a]Twenty-five records were missing; [b]McNemar's Test.

Alteration of behaviors associated with IAP exposure comparing the pre and post intervention periods are additionally shown in Table 2. These behaviors were compared to assess the health education component of the intervention. Using portable open fires to warm the pot tea, a tradition kept for hundreds of years in this region, decreased from 35.6% before the intervention to 33.7% after the intervention ($p = 0.01$). During the 2-year intervention, more households chose to buy and equip fans in kitchens to exhaust cooking smoke, and this use increased from 1.0% to 5.6% ($p < 0.01$) among enrolled participants. Furthermore, the percentage of participants who never opened windows during winter reduced from 16.0% to 11.1% over the course of the intervention period ($p = 0.048$; Table 2).

3.3. Improvement of Indoor Air Quality

Households for IAP sampling were selected from the 371 households, considering house area, type of fuel used, stove features, and ventilation on the basis of a pilot study. Classifying by quantiles in accordance with questionnaire data of 413 women, the sample distribution of household areas was similar between the total population and the selected 8 households. Overall, the characteristics of the households were similar comparing the total study population and those households selected for sampling (Table 3), and in addition sociodemographic characteristics including income and education level were similar among women in households with and without IAP monitoring.

A comparison of PM_4 and CO levels before and after the intervention stratified by room type is shown in Table 4 and Table 5. The average concentration of PM_4 decreased by about 70%, from 774 to 223 μg/m^3, in the kitchen following implementation of the intervention, though this effect was marginally significant ($p = 0.08$) due to the small sample size. Conversely, there was no observed decrease in the average bedroom concentrations of PM_4 following the intervention (pre 135 μg/m^{3}vs. post 176 μg/m^3; $p = 0.20$), whereas the average PM_4 levels of the bedroom and kitchen combined were decreased by about 56% following the intervention (pre 455 μg/m^3 vs. post 200 μg/m^3; $p = 0.46$). The concentration of PM_4 in the bedrooms was lower than that in the kitchen both before and after stove improvement (Table 4).

The average concentrations of CO in both the kitchen and bedroom were reduced following the intervention (Table 5). Specifically, CO levels decreased in the kitchen from 3.81 ppm to 3.00 ppm, while concentrations reduced from 2.99 ppm to 2.80 ppm in the bedroom. For both rooms combined, CO levels changed slightly, declining from 3.40 ppm to 2.90 ppm, although none of these reductions of CO were statistically significant (Table 5).

3.4. Changes of Health Indicators of Women

Changes in lung function following the intervention are reported in Table 6. Spirometry of women showed that FEV1 increased from 2.33 ± 0.54 L/s to 2.46 ± 0.44 L/s and FVC increased from 2.75 ± 0.70 L to 2.84 ± 0.56 L comparing the levels before and after the intervention. The percentage of predicted value of FEV1 (pre-inter- vention 86.7% ± 18.8% vs. post-intervention 93.9% ± 14.5%) and FVC (pre-intervention 89.9% ± 21.8% vs. post-intervention 95.0% ± 16.3%) improved to some degree after the intervention, but all the parameters of lung function assessment did not show significant change. In the 49 participants who had lung function assessed, the prevalence of self-reported eye and respiratory symptoms that occurred in the past 3 months from the post-intervention survey decreased for all symptoms examined compared to the reported prevalence from baseline, including tearing, sore eyes, red eyes, runny nose, nose congestion, continuous sneeze, phlegm, and fever (Table 7).

Among the total study population, a similar reduction in the percent prevalence of all examined symptoms was apparent following the intervention compared to baseline (Table 7). Significant reductions were observed for the prevalence of tearing, which decreased from 5.6% to 0%, sore eyes which decreased from 3.6% to 1.5%, and for red eyes which decreased from 3.9% to 1.0%. Further, there were significant reductions in the percentage of participants reporting phlegm (pre 6.8% vs. post 3.4%; $p = 0.01$) and fever (pre 5.3% vs. post 1.7%; $p < 0.01$; Table 7).

4. DISCUSSION

The results from this intervention suggest that stove improvement and associated behavioral changes related to fuel use, such as opening windows while cooking, may be effective in mitigating the negative health consequences associated with IAP. Specifically, we observed a reduction in average PM_4 levels, particularly in the kitchen area, and in the prevalence of symptoms associated with IAP following implementation of the stove improvement and behavioral intervention. Further, FEV1 favorably increased following the intervention compared to baseline, although this effect was not statistically significant. The households selected for IAP assessment were shown to be representative of the larger study population with respect to home area, fuel type used, and stove features and ventilation practices, suggesting that our findings may be generalizable to the overall study population. While the average bedroom concentration of PM_4 was slightly higher after stove improvement, this is potentially attributed to the small sample size of monitored households and/or the limited duration of each measurement period (i.e., 24 hours).

Table 3. Selected characteristics of households for questionnaire survey and for IAP monitoring (%).

	Households for questionnaire survey (n = 413)	Households for IAP monitoring (n = 8)
House area (m^2)[a]		
<60	14.6	12.5
60 - 79	33.0	25.0
80 - 99	22.3	25.0
100~	30.1	37.5
Biomass use	100.0	100.0
Stoves without smoke door[b]	88.1	75.0
Chimney lower than eave[c]	77.9	87.5

[a]Household areas were classified by quantiles based on the data of questionnaire survey. One record was missed; [b]Two records were missing; [c]Thirty-three records were missing.

Table 4. Comparison of PM_4 levels before and after intervention ($\mu g/m^3$).

Sampling spots	Time	N[a]	Mean	SD	Median	p-value[b]
Kitchen	Before	8	774	756	568	0.08
	After	8	223	104	197	
Bedroom	Before	8	135	149	100	0.20
	After	8	176	28	163	
Total	Before	8	455	621	183	0.46
	After	8	200	77	163	

[a]Sample size of households for IAP monitoring; [b]Wilcoxon signed rank test comparing data before and after intervention.

Table 5. Comparison of CO levels before and after intervention (ppm).

Sampling spots	Time	N[a]	Mean	SD	Median	p-value[b]
Kitchen	Before	8	3.81	3.25	5.26	0.94
	After	8	3.00	1.57	3.05	
Bedroom	Before	8	2.99	2.79	2.31	0.84
	After	8	2.80	1.77	2.53	
Total	Before	8	3.40	2.96	3.81	0.72
	After	8	2.90	1.62	2.86	

[a]Sample size of households for IAP monitoring; [b]Wilcoxon signed rank test comparing data before and after intervention.

Table 6. Changes in lung function before and after intervention (n = 49).

Variables	Observed		Percentage predicted (%)		Mean changes[a]	t[b]	p-value[b]
	Before	After	Before	After			
VC (L)	3.20 ± 0.57	3.10 ± 0.51	104.5 ± 18.3	103.6 ± 16.0	−0.9	0.32	0.75
FVC (L)	2.75 ± 0.70	2.84 ± 0.56	89.9 ± 21.8	95.0 ± 16.3	5.1	1.11	0.27
FEV1 (L/s)	2.33 ± 0.54	2.46 ± 0.44	86.7 ± 18.8	93.9 ± 14.5	7.1	1.91	0.06
FEF 25% - 75% (L/s)	2.87 ± 0.85	2.71 ± 0.56	87.9 ± 26.9	84.5 ± 17.2	−3.5	−1.61	0.12

[a]Mean changes based on percentage predicted value of lung function (the values after intervention were subtracted by the values before intervention); [b]Adjusted using a linear mixed model for cooking years up to date of survey, using open fire, no fans in kitchens, location of kitchen and never opening windows when cooking.

Table 7. Prevalence rates of symptoms before and after intervention among spirometry participants and total subjects (%).

Symptoms	Spirometry participants		(n = 49)	Total subjects		(n = 413)
	Before	After	p-value[a]	Before	After	p-value[b]
Tearing	14.3	2.0	0.01	5.6	0	---
Sore eyes	8.2	4.1	0.41	3.6	1.5	0.04
Red eyes	12.2	2.0	0.06	3.9	1.0	0.01
Runny nose	2.0	0	-	2.2	1.5	0.70
Nose congestion	4.1	0	-	2.7	1.0	0.11
Continuous sneeze	6.1	0	-	2.4	1.2	0.34
Phlegm	8.2	4.1	0.41	6.8	3.4	0.01
Fever	10.2	2.0	0.10	5.3	1.7	<0.01

[a]McNemar's Test; [b]Adjusted for age, cooking years up to date of survey, using open fire, location of kitchen and never open windows when cooking using logistic regression.

Since few nations have indoor air quality standards for PM_4, we can only compare the result with the limit for PM_{10} in domestic air. In China, the national standard for indoor air quality dictates that the average concentration of PM_{10} should not exceed 150 μg/m³ [19] . While stove improvement was demonstrated to reduce PM_4 levels, the concentration of PM_4 in indoor air was still higher than this standard for PM_{10}. Using solid fuels was the root of IAP in rural households for the limit of economic status. It would be difficult to reduce the indoor air pollutants to very low levels unless clean fuels were prevalent in rural households. While our results indicated significant behavioral modifications concerning fuel use habits, some lifestyles, such as using portable open fires to heat tea, were still popular after the intervention. Consequently, this may have contributed to the relatively high levels of PM_4 even after stove improvement in some households.

Changes in CO levels generally suggested a mild trend of decline, but the measured concentrations were lower than expected. Specifically, 24 hour average concentrations of CO before and after intervention were consistently below current standards and guidelines (i.e., WHO guideline values of 5.6 ppm/7 mg/m^3 for 24 h exposures), and in some cases CO concentrations were close to the detection limits of the diffusion tubes. Since these were 24-h concentrations, concentrations may have been higher during cooking or when doors and windows were closed at night, but this was not observable in our data [16] . Furthermore, about 83% of the surveyed women use to open windows in the winter and 84% opened kitchen windows for ventilation when cooking even before the behavior intervention; therefore, generated CO may have diffused to others rooms or outdoors. For both CO and PM_4, the percent reduction in levels before and after the intervention in the field study were lower compared to previously conducted controlled tests [16] . This may primarily be due to other factors associated with levels of these pollutants, such as individual cooking practices and ventilation patterns, and also due to the fact that the controlled tests were conducted only during periods when the stove was burning.

In our study, socioeconomic differences, individual biological variations, and some potential confounders that might affect lung function among populations were controlled because we compared the changes in the same people before and after intervention. In addition, we transformed the observed values to percentages of predicted values in order to control for age and possible changes of body composition during the 2-year study period, and used linear mixed models to assess the improvement of lung function parameters, adjusting for other potential confounders. In some follow-up studies on lung function, spirometry results may have been affected by participants who repeated the test several times and developed experience and skills to perform better on the respiratory function test [20] . However, this was unlikely to have influenced the results in our study given the relatively long interval between the first and second lung function assessments. Data in this study suggested the possible improvement of FEV1 in women to some degree after IAP mitigation. However, this finding should be interpreted cautiously since other parameters of lung function did not show similar improvement. The post-interven- tion examination of lung function was conducted roughly 5 months after the large-scale stove intervention. It would be unlikely for these parameters to improve right after intervention measures had been taken. Follow-up studies are warranted in order to evaluate the long term effect of IAP mitigation on lung function.

On the other hand, we observed significant decreases in the prevalence rates of some eye and respiratory symptoms, as well as for fever, following the intervention. In other community studies carried out in developing countries, investigators also found that biomass exposure was consistently associated with chronic respiratory symptoms although effects on lung function were variable or small [5] [20] [21] . These studies suggested that symptoms were more sensitive to variation of IAP levels. Since accurate disease records were not available in this region, the symptoms survey was alternatively used to measure the variation in health status. Given that eye and respiratory symptoms reflect stimulation of the mucosas and aggravation or attack of diseases, the decreases in prevalence of these symptoms provide a potential indication of improvement in health status following the intervention and reduction in IAP levels. Our results also indicated similarly decreasing trends in the prevalence of symptoms in the 49 women who took part in the lung function test, which suggested health improvement after intervention similar to the overall study population. These women also had relatively higher prevalence rates of all symptoms except runny nose during the baseline period and more significant reduction in the post-interven- tion interview. This might be attributed to the fact that people in poorer health condition were more likely to participate in the IAP mitigation activities and would get more improvement in health. However, we note that the study personnel and participants in the study were not blinded to the intervention status, and therefore there is a possibility of biased reporting for the subjective physical symptoms. In addition, it is possible that some reported symptoms may have been due to other causes aside from the effects of IAP, although participants were asked about the prevalence of symptoms over a three month period before the interview, which provided some determination of whether the symptoms were chronic in nature rather than from acute respiratory infections.

There were some additional limitations in this study. First, combustion of solid fuels was the source of indoor air pollutants in rural regions of China. However, it would not be feasible to implement cleaner energy (e.g. natural gas and electricity) given the poor economic conditions among the rural households and undeveloped infrastructure. The primary goal of this intervention was therefore to inform the public of the adverse effects of IAP from biomass and assist them in replacing traditional stoves. The appearance, usage, and maintenance of the new stoves were similar to the traditional ones and thus the effect of the intervention is likely to be sustainable. Second, the exposure assessment conducted in our study utilized area sampling for a relatively short time period (i.e., 24 hours) rather than using personal monitors for the study subjects. This approach provided a general indication of the levels of IAP in the monitored households, but might not capture the variability in exposures and could either underestimate or overestimate personal exposure depending on the placement of the monitors in relation to the participants' usual activities. Third, our study was relatively small in terms of the number of monitored households and subjects included in lung function testing, and the timeline of the project required a shorter than ideal interval between the baseline and follow-up periods and therefore follow-up studies will be needed. Finally, some risk behaviors practiced for generations were difficult to change in the one- or two-year intervention period.

5. CONCLUSION

In conclusion, we have shown that an intervention consisting of stove improvement and behavioral modifications related to fuel use may be an effective and viable method to mitigate the deleterious health effects of IAP in developing regions. Specifically, our results suggest that such an intervention may reduce levels of PM_4 and physical symptoms associated with IAP exposure, and in addition may improve some measures of lung function, although the effects reported in our study are not statistically significant due to the relatively small size of our study and the symptoms assessed are based on self-report. Follow-up studies with a longer assessment period between the pre and post intervention periods are clearly needed to properly evaluate changes in lung function associated with these stove improvements, as well as the long-term effects of the intervention.

ACKNOWLEDGEMENTS

The authors would like to express special gratitude to all the study subjects in the study and the personnel in the area of Gansu Province, China, which have supported the study. Funding for this work was provided by the World Bank through financial support from the Energy Sector Management Assistance Program (ESMAP), Department for International Development, UK (DFID) and Swedish International Development Agency (SIDA) and also partly supported by Fogarty training grants D43TW 008323 and D43TW 007864-01 from the National Institutes of Health.

REFERENCES

1 World Health Organization (2013) World Health Statistics. http://www.who.int/gho/publications/world_health_statistics/2013/en/

2 World Health Organization (2007) Indoor Air Pollution: National Burden of Disease Estimates. World Health Organization, Geneva.http://www.who.int/indoorair/publications/indoor_air_national_burden_estimate_revised.pdf

3 Lim, S., Vos, T., Flaxman, A.D., Danaei, G., Shibuya, K., Adair-Rohani, H., et al. (2012) A Comparative Risk Assessment of Burden of Disease and Injury Attributable to 67 Risk Factors and Risk Factor Clusters in 21 Regions, 1990-2010: A Systematic Analysis for the Global Burden of Disease Study 2010. The Lancet, 380, 2224-2260.

4 Kurmi, O.P., Semple, S., Simkhada, P., Smith, W.C. and Ayres, J.G. (2010) COPD and Chronic Bronchitis Risk of Indoor Air Pollution from Solid Fuel: A Systematic Review and Meta-Analysis. Thorax, 65, 221-228.

5 Regalado, J., Perez-Padilla, R., Sansores, R., Paramo Ramirez, J.I., Brauer, M., Pare, P., et al. (2006) The Effect of Biomass Burning on Respiratory Symptoms and Lung Function in Rural Mexican Women. American Journal of Respiratory and Critical Care Medicine, 174, 901-905.

6 Díaz, E., Smith-Sivertsen, T., Pope, D., Lie, R.T., Díaz, A., McCracken, J., et al. (2007) Eye Discomfort, Headache and Back Pain among Mayan Guatemalan Women Taking Part in a Randomised Stove Intervention Trial. Journal of Epidemiology and Community Health, 61, 74-79.

7 Liu, S., Zhou, Y., Wang, X., Wang, D., Lu, J., Zheng, J., et al. (2007) Biomass Fuels Are the Probable Risk Factor for Chronic Obstructive Pulmonary Disease in Rural South China. Thorax, 62, 889-897.

8 Mishra, V. (2003) Effect of Indoor Air Pollution from Biomass Combustion on Prevalence of Asthma in the Elderly. Environmental Health Perspectives, 111, 71-78.

9 Mestl, H.E., Aunan, K., Seip, H.M., Wang, S., Zhao, Y. and Zhang, D. (2007) Urban and Rural Exposure to Indoor Air Pollution from Domestic Biomass and Coal Burning across China. Science of the Total Environment, 377, 12-26.

10 World Health Organization (2009) Global Health Risks: Mortality and Burden of Disease Attributable to Selected Major Risks.

http://www.who.int/healthinfo/global_burden_disease/GlobalHealthRisks_report_full.pdf

11 Jin, Y., Ma, X., Chen, X., Cheng, Y., Baris, E. and Ezzati, M. (2006) Exposure to Indoor Air Pollution from Household Energy Use in Rural China: The Interactions of Technology, Behavior, and Knowledge in Health Risk Management. Social Science Medicine, 62, 3161-3176.

12 He, G., Ying, B., Liu, J., Gao, S., Shen, S., Balakrishnan, K., et al. (2005) Patterns of Household Concentrations of Multiple Indoor Air Pollutants in China. Environmental Science Technology, 39, 991-998.

13 Chapman, R.S., He, X., Blair, A.E. and Lan, Q. (2005) Improvement in Household Stoves and Risk of Chronic Obstructive Pulmonary Disease in Xuanwei, China: Retrospective Cohort Study. British Medical Journal, 331, 1050.

14 Lan, Q., Chapman, R.S., Schreinemachers, D.M., Tian, L. and He, X. (2002) Household Stove Improvement and Risk of Lung Cancer in Xuanwei, China. Journal of the National Cancer Institute, 94, 826-835.

15 Shen, M., Chapman, R.S., Vermeulen, R., Tian, L., Zheng, T., Chen, B.E., et al. (2009) Coal Use, Stove Improvement, and Adult Pneumonia Mortality in Xuanwei, China: A Retrospective Cohort Study. Environmental Health Perspectives, 117, 261-266.

16 Zhou, Z., Jin, Y., Liu, F., Cheng, Y., Liu, J., Kang, J., et al. (2006) Community Effectiveness of Stove and Health Education Interventions for Reducing Exposure to Indoor Air Pollution from Solid Fuels in Four Chinese Provinces. Environmental Research Letters, 1, 1-12.

17 Jin, Y., Zhou, Z., He, G., Wei, H., Liu, J., Liu, F., et al. (2005) Geographical, Spatial, and Temporal Distributions of Multiple

Indoor Air Pollutants in Four Chinese Provinces. Environmental Science Technology, 39, 9431-9439.

18 Mu, K. and Liu, S. (1990) National Compilation of Normal Value of Lung Function. Peking Union Medical College, Beijing Medical University Joint Publishing House, Beijing.

19 Ministry of Health (2002) Indoor Air Quality Standard Ministry of Health, China.http:// www.moh.gov. cn/publicfiles// business/ htmlfiles/ zwgkzt/pwsbz/index.htm

20 Smith-Sivertsen, T., Dıaz, E., Pope, D., Lie, R.T., Dıaz, A., McCracken, J., et al. (2009) Effect of Reducing Indoor Air Pollution on Women's Respiratory Symptoms and Lung Function: The RESPIRE Randomized Trial, Guatemala. American Journal of Epidemiology, 170, 211-220.

21 Rinne, S., Rodas, E., Bender, B., Rinne, M., Simpson, J., Galer-Unti, R., et al. (2005) Relationship of Pulmonary Function among Women and Children to Indoor Air Pollution from Biomass Use in Rural Ecuador. Respiratory Medicine, 100, 1208-1215.

CHAPTER 7

Indoor Air Quality Measurement with the Installation of a Rooftop Turbine Ventilator

Jason Lien, Noor Ahmed

School of Mechanical and Manufacturing Engineering, University of New South Wales, Kensington, Australia.

ABSTRACT

The present paper presents a numerical analysis of the difference in comfort level inside a room of a residential building when roof top turbine ventilator is installed. This analysis simulates various comfort factors which includes the indoor air movement, room temperature, Predicted Mean Vote (PMV) and Predicted Percentage of Dissatisfied (PPD). Various test cases of ventilator exhaust rate were examined. The results showed that general comfort satisfying international standards in building can be achieved. This study also presents a qualitative and quantitative study of indoor air temperature and overall indoor air flow pattern. A promising conclusion that can be drawn from this study is that wind driven ventilators can play an important role in the design of a cost effective and energy efficient ventilation system inside a building.

KEYWORDS

Indoor Air Quality; Rooftop Turbine Ventilator; Computational Analysis

1. INTRODUCTION

Inadequate ventilation in a building can result in problems with moisture, unpleasant smell, lack of oxygen, and unacceptable content of poisons gases such as CO. Contaminants such as formaldehyde or radon can also accumulate in poorly ventilated homes, causing health problems. However, since resistance to pollutant varies from person to person, it is hard to accurately quantify the impact of ventilation on human beings. Various methods have been proposed based on the consideration of factors such as direct medical costs and lost earnings due to major illness as well as increased employee sickness days and lost productivities while on the job. From such studies, the productivity loss in the United States of America attributed directly to indoor air quality (IAQ) has been estimated to be approximately 14 minutes per day [1] or 1 to 1.5 billion of euro per year or equivalent of 250 to 350 euro per inhabitant in Norway [2]. Other studies suggest that the potential economic impact of indoor air pollution on a country, such as on Finland, can run into tens of billions of dollars per calendar year [3].

The last century has seen tremendous progress in technological development that is underpinning the progress of the modern day human civilization. The focus on these developments has been placed on energy efficiency and sustainable buildings for most governments around the world. Moreover, recent price hike in fossil and energy accelerates people's awareness and attention in focusing the use of more sustainable solutions other than relying on fossil fuel. As a result, natural ventilation has become an increasingly attractive way of reducing energy consumption and cost in recent decades.

Engineers have exhausted ideas to evaluate the most cost effectiveness measures to achieve a renewable and sustainable goal. The Aerodynamic Research team at the University of New South Wales over the last decade has been actively applying novel aerodynamic concepts [4- 12] to harness the benefit of renewable energy by developing various methodologies and techniques [13-26] for cost effective environmental friendly devices.

Most of the measures applied on buildings sector have aimed to improve performance across all the metrics that include: saving building energy consumption, water efficiency, green house gas emissions reduction, indoor environmental quality improvement and stewardship of resources and sensitivity of their impacts [27]. Under such concern, assisted natural ventilation devices such as turbine ventilators became more favorable and viable solution in providing suitable indoor environment for many building designs.

Typical wind driven turbine ventilators as shown in **Figure 1**, consist of a base and a top in the form of a rotor, which acting like a turbine, extracts mechanical work from the wind to produce its rotation. Using of these devices can greatly assist to improve the indoor air quality when strategically placed [28].

Experimental studies and CFD models have been applied to evaluate various indoor environments, such as apartments, offices, classrooms and factories [29-32]. The fundamental aerodynamic behaviour of rotating ventilator has been analysed in both numerically and experimentally [33]. During the design phase, the process often involves determination of the optimal conditions of ventilation system, such as air velocity and temperature, location of air inlet and exhaust outlet. Hence, computational fluid dynamics (CFD) simulation method would become an efficient and economical tool to control and examined indoor environment [34].

Figure 1. Rooftop turbine ventilators installations.

The present paper extends the previous studies of Lien and Ahmed [33], focusing on the comfort level difference inside a typical residual house room when roof top turbine ventilator is installed. The main comfort factors studied includes the indoor air movement, room temperature, Predicted Mean Vote (PMV) and Predicted Percentage of Dissatisfied (PPD).

2. NUMERICAL MODEL

2.1. The Computation Code

Commercial CFD packages are frequently used for indoor environment simulations and calculations [35-37].

In the present work the latest version of FLUENT is used. The equations for the conservation of mass, momentum and turbulence scales are solved in FLUENT using the control volume method in a three-dimensional bodycoordinate system. Recognising the results of CFD studies largely depend on the meshing quality, turbulence models, boundary conditions and difference schemes, a brief outline of them as used in the present modelling is summarised here.

2.2. Boundary Conditions

A model house living room with dimension of 5 m long × 4 m wide × 3 m height was simulated in this study and is shown in **Figure 2**(a). A ventilator duct was installed on top ceiling (11 cm diameter). An air inlet vent (0.5 × 0.5 m) was also installed on the left wall to supply fresh air. The air supply temperature is set to 23°C to simulate a typical Sydney day temperature. This simulation models the natural ventilation mechanism hence no inlet velocity property was assigned. The ventilator exhaust rate was controlled by using CFD software to observe the indoor comfort variables change. Eight testing points are employed to collect steady state condition. The locations of these test points are shown in **Figure 2**(b). The final total number of grids used in the present simulation was determined to be approximately 800,000. Mesh quality was checked to ensure the skewness of the three dimensional grids is less than 0.8.

The accuracy of CFD simulations relies on the appropriate setting of physical boundary conditions. The controlled boundary conditions of the present work includes: the outside ambient condition, indoor heat sources generation, ventilator exhaust rate (air outlet), vent air inlet, humidity and radiation. The amount of heat generated by occupants and electrical appliances are represented by heat fluxes. Assumption of these heat fluxes are applied, i.e. the occupant metabolic rate is 70 W/m^2, each lamps generates 34 W of heat and 300 W for the Television unit. Atmospheric pressure was applied on the inlet and outlet condition. The air humidity set to at 50%.

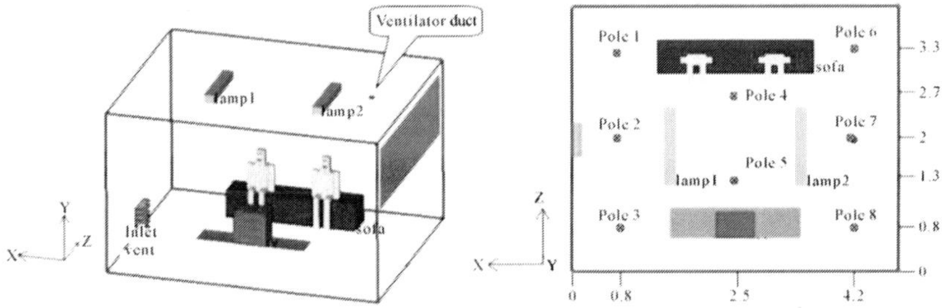

Figure 2. (a) Schematic of the model room; (b) Testing pole locations.

2.3. The Choice of Turbulence Model

The standard k-e model has been widely used in the industrial sector for its effectiveness in various engineering applications. For many cases, the choice of standard k-e was a good compromise except for natural ventilation with significant indoor thermal loads [38]. However, the standard k-e turbulence model has suffered from deficiencies for areas with very low velocities, particularly near wall regions. A basic characteristic of the RNG k-e turbulence model is that the analytically derived differential formula accounts for low Reynold number effects. Hence for near wall regions, the RNG k-e turbulence model are expected to be more effective [31].

In the present work, RNG k-e based model is used to predict airflow velocity and temperatures in the model house room.

2.4. Standard Regulation Criteria

Following ISO 7730 [38], the parameters which defines the thermal environment are temperature, humidity, air velocity, and personal parameters such as clothing with activity level. For a comfort indoor environment (80% acceptance), the benchmark values are summarised below:

- Local indoor air velocity < 0.25 m/s.
- Desired range of room temperature 24°C - 26°C.
- Predicted Mean Vote (PMV) margin –0.5 - 0.5.
- Predicted Percentage of Dissatisfied (PPD) ≤ 10%.

These comfort variables are monitored at all testing pole and compared with the benchmark values.

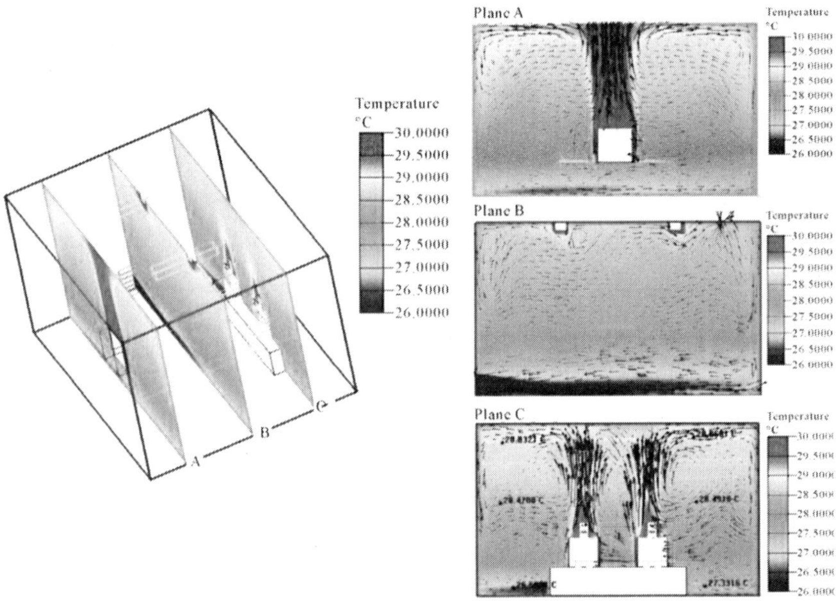

Figure 3. Airflow pattern and room temperature contour plot.

3. RESULTS AND OBSERVATIONS

Four different ventilation exhaust rate of were 1 m/s, 1.5 m/s, 2 m/s, 3 m/s were simulated. For the purpose of observation, the exhaust rate was set at 1.5 m/s and the various flow pattern and temperature contour were plotted at three cut planes (z = 0.8 m, 2 m, 3.3 m) along the XY and YZ direction

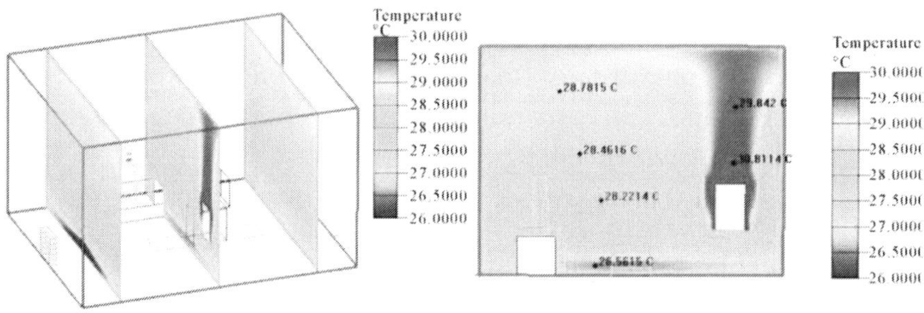

Figure 4. Temperature contour plot in YZ plane.

3.1. Flow Field and Temperature Observations

The ventilated flows at three cut-planes were calculated and shown in Figures 3 and 4. The flow pattern results agree well with the experimental and numerical work by Stamou and Katsiris, 2006 [31] where an office room was modelled. Four distinct flows regions can be characterised as follows:

1) The floor layer and incoming flow region The incoming flow (23°C) through the vent is cooler than the room temperature (approximately 26°C to 30°C). A floor layer is formed as the cool air enters the room via the inlet, bends downwards due to buoyancy force.

2) The buoyant plumes region Large buoyant vertical plumes are formed, see Figures 3 and 4. These are originated from the heat sources; the occupant, the television unit, and the lights.

3) The ceiling layer and the outlet flow region The plumes from the heat sources travelled upwards until struck the ceiling and flowing along towards all directions. This phenomenon is shown in Figures 3 and 4 (Cut planes A and C).

4) The exit flow and re-circulation region The ventilator draws exhaust flow to exit from the ventilation duct and this creates a re-circulation region in the centre of the room. Other re-circulation regions were observed near the heat sources, largely caused by the buoyant plumes. The temperature profiles generally increases with the room vertical height. Higher temperature is observed in the region of the buoyant jets and the ceiling layer. The room temperature, except the buoyant plumes region, can be divided into three main zones: the cooler, the intermediate and the warmer. The results also indicate that the horizontal temperature variation is considerably less than the vertical profiles.

3.2. Results at the Testing Pole

After analysed the flow field, the comfort level in the occupied zone is evaluated. Environment variables i.e. air temperature, air velocity and mean radiant temperature were calculated at 8 testing poles as shown in **Figure 2**. The data were collected from the floor up to a height of 2.5 meters under steady-state condition. The variations of room temperature, air velocity and PMV with height were plotted at each pole as shown in Figures 5-7 respectively. The exhaust rate is defined at 1.5 m/s.

The temperature plots indicated that the temperature showed an increasing trend with the height. The overall room temperature varied from 26.3°C to 28.8°C, except a low region of 24.7°C, which was near the air inlet vent of 23°C. Due to the fact that the overall flow pattern strongly depends on temperature change, the temperature gradient needs to be observed. From the results, it can be concluded that all temperature curves showed the gradient with three stages:

1) From floor to approximately 0.5 m At this range the temperature changed relatively rapid due to the heat convection between the bottom ground cold air and the hot air inside the room.

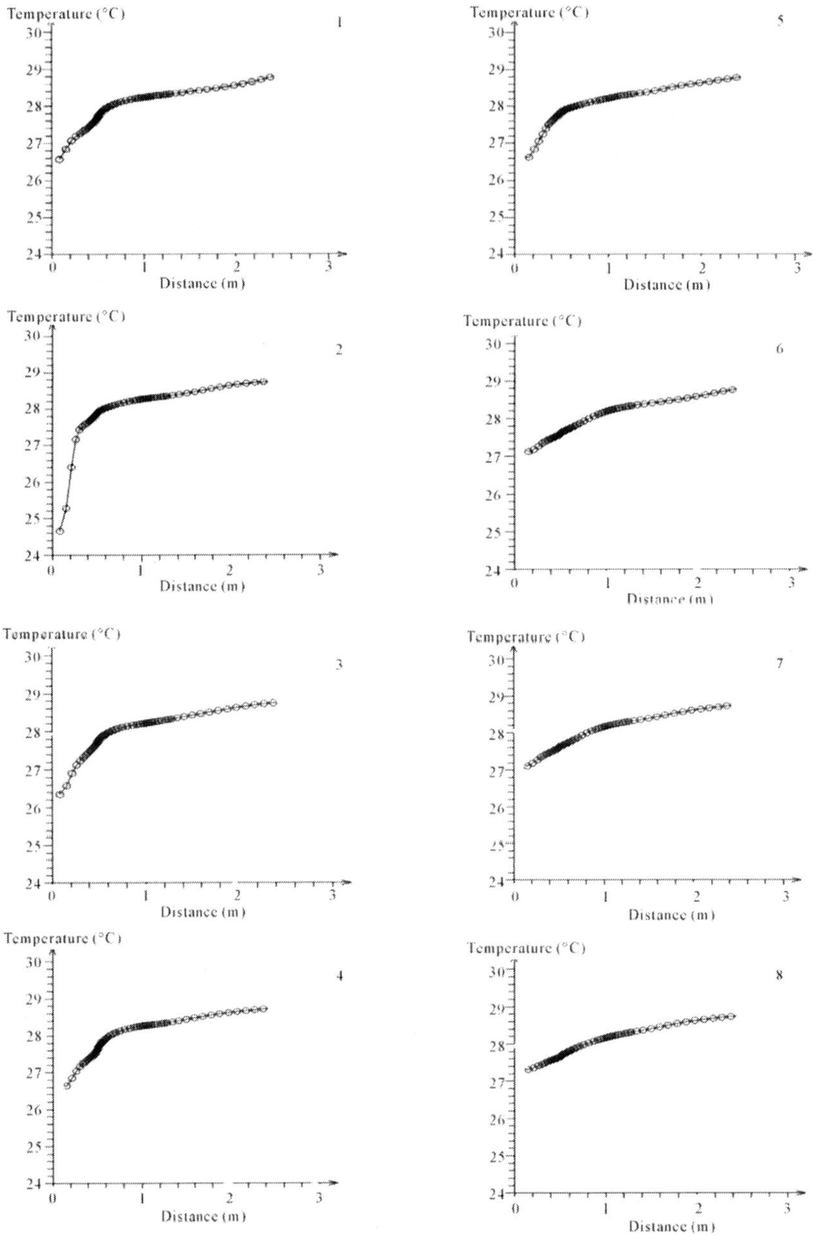

Figure 5. Temperature variation for testing poles 1-8.

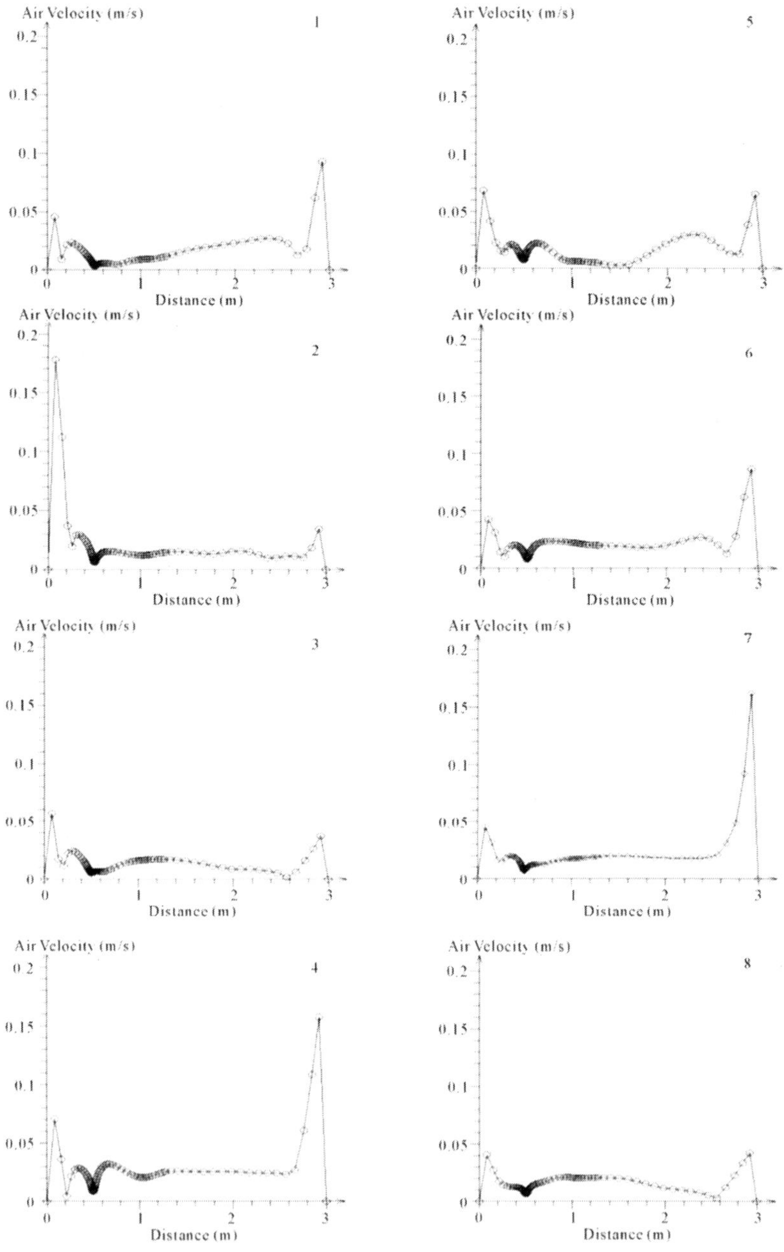

Figure 6. Air velocity variation for testing poles 1-8.

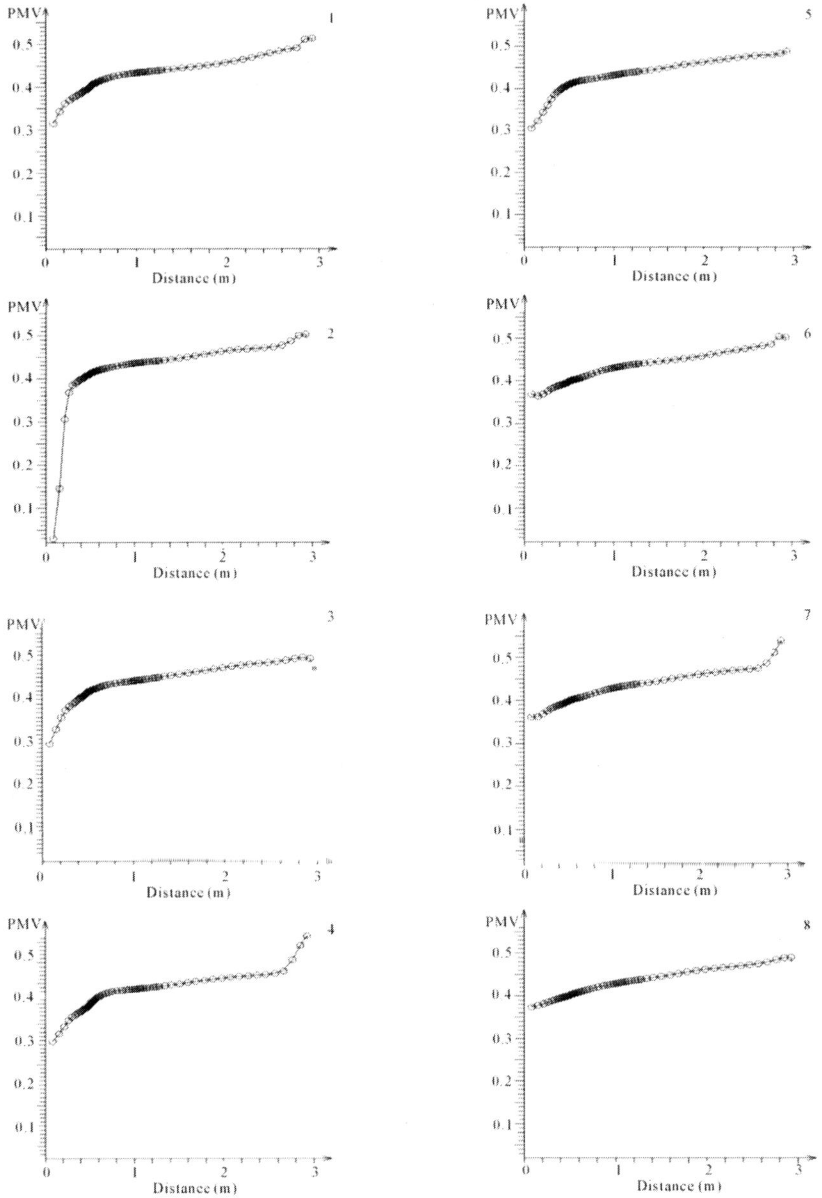

Figure 7. PMV variation for testing poles 1-8.

2) From 0.5 to 1.4 m in height The decreasing change of temperature gradient was observed. This indicated that the air temperature in the middle layer of the room varied very small; finally.

3) From 1.4 to 2.5 m in height The temperature gradient in this section is relatively smaller compare to the previous sections. The main flow driving force in this section is the buoyancy effect.

The air velocity variation plots indicate that peak air movement occurred near both ground and ceiling regions. The region between these two peaks has air velocity less than 0.05 m/s. It is noted that, areas with high air movement is at the bottom region of pole 2 and the top region of pole 7 which correspond well with the boundary conditions as they were located near the air inlet and outlet. The minimum air velocity movement region was around 0.5 m except for pole 4. This coincides with the observations of the flow pattern described in the previous section as the interaction of two bottom layer flow in opposite directions.

The PMV value largely depend on temperature, hence the trend of PMV plot as shown in **Figure 6** is similar to temperature plots. The PMV values of all 8 testing poles within 2.5 m height are less than 0.5.

After observing the trend of the interested comfort variables, the mean values of variable are recalculated for the regions of 0 to 2.5 m height with different ventilator exhaust rate (1, 1.5, 2 and 3 m/s). The results are shown in tabulated form where the temperature and air velocity are shown in **Table 1**and the PMV and PPD values are shown in **Table 2**.

4. DISCUSSION OF INDOOR COMFORT LEVEL

This study focused on the study of indoor air quality and comfort level when the turbine ventilator is installed. Four major comfort qualities: air velocity, temperature, PMV and PPD are evaluated and compared with ISO Standard 7730. Each of these aspects is discussed in this section.

4.1. Air Movement

The simulated results of the four exhaust rate indicate that air movement velocity at all test poles are within the range of 0.04 - 0.21 m/s. The highest air velocity was observed at the region near the inlet vent and the lowest one located at pole 8. According to ISO Standard 7730, the modelled air velocities are well within the comfort criteria of 0.25 m/s.

4.2. Temperature

The results in **Table 1** indicate that the mean temperature at all test poles reduced from 27.71°C to 27.26°C as the ventilator exhaust rate increased from 1 m/s to 3 m/s. However this is slightly higher than the desired range of temperature of 24°C - 26°C as indicated by ISO Standard 7730.

4.3. Predicted Mean Vote (PMV) and Predicted Percentage Dissatisfied (PPD)

The acceptance to climate change for each individual is different, hence the evaluation process for comfort level is a very subjective task [38]. During the past decade, efforts by researchers have been spent to develop the international standards for maintaining comfortable indoor environment. The common used standards are ASHRAE [39] and ISO 7730 [38]. In both standard, the PMV model uses a steady-state heat balance for the human body and it is related to 7 thermal sensation levels of the occupants who exposed to a certain environment. The PMV levels usually range between −3 and +3, where the negative levels stand for cold feeling and the positive values for hot feeling, 0 being a neutral or optimum level. PPD value indicates the variance in the thermal sensation of the occupants exposed to the same condition with dissatisfaction or discomfort. The ISO standard defines that an acceptable indoor thermal environment for general comfort is: $-0.5 < \text{PMV} < 0.5$ and $\text{PPD} < 10\%$ for a space where occupants have activity levels with metabolic rates between 1.0 and 1.3 and the thermal insulation worn is between 0.5 and 1.0 clo.

Table 1. Calculated air velocity and temperature for 0 - 2.5 m vertical height.

Ventilator exhaust rate:	V = 1 m/s				V = 1.5 m/s			
	Temperature (°C)			Velocity (m/s)	Temperature (°C)			Velocity (m/s)
Testing sector:	Min	Max	Mean	V_{max}	Min	Max	Mean	V_{max}
Pole 1	26.70	28.90	27.80	0.04	26.60	28.80	27.70	0.05
Pole 2	24.70	28.70	26.70	0.17	24.70	28.70	26.70	0.18
Pole 3	26.40	28.80	27.60	0.05	26.30	28.70	27.50	0.07
Pole 4	26.80	28.70	27.75	0.07	26.60	28.60	27.60	0.07
Pole 5	26.70	28.80	27.75	0.07	26.70	28.70	27.70	0.07
Pole 6	27.20	28.80	28.00	0.04	27.10	28.60	27.85	0.04
Pole 7	27.20	28.80	8.00	0.04	27.10	28.70	27.90	0.05
Pole 8	27.40	28.70	28.05	0.04	27.30	28.60	27.95	0.04

T_{avg} 27.71; T_{avg} 27.61.

Ventilator exhaust rate:	V = 2 m/s				V = 3 m/s			
	Temperature (°C)			Velocity (m/s)	Temperature(°C)			Velocity (m/s)
Testing sector:	Min	Max	Mean	V_{max}	Min	Max	Mean	V_{max}
Pole 1	26.40	28.70	27.55	0.05	26.20	28.60	27.40	0.06
Pole 2	24.50	28.60	26.55	0.19	24.30	28.40	26.35	0.21
Pole 3	26.20	28.50	27.35	0.06	26.00	28.50	27.25	0.06
Pole 4	26.40	28.50	27.45	0.08	26.20	28.40	27.30	0.09
Pole 5	26.30	28.60	27.45	0.08	26.10	28.50	27.30	0.09
Pole 6	26.90	28.60	27.75	0.05	26.70	28.50	27.60	0.05
Pole 7	26.80	28.60	27.70	0.05	26.20	28.50	27.35	0.06
Pole 8	26.90	28.50	27.70	0.05	26.70	28.40	27.55	0.05

T_{avg} 27.44; T_{avg} 7.26.

Table 2. Calculated PMV and PPD for 0 - 2.5 m vertical height.

Ventilator exhaust rate:	V = 1 m/s					
	PMV			PPD		
Testing sector:	Min	Max	Mean	Min	Max	Mean
Pole 1	0.33	0.48	0.40	7.21	9.90	8.55
Pole 2	0.05	0.48	0.26	5.14	9.66	7.40
Pole 3	0.30	0.48	0.39	6.85	9.80	8.33
Pole 4	0.32	0.47	0.39	7.19	9.70	8.45
Pole 5	0.32	0.48	0.40	7.06	9.73	8.40
Pole 6	0.37	0.48	0.43	7.85	9.80	8.83
Pole 7	0.37	0.47	0.42	7.82	9.70	8.76
Pole 8	0.38	0.48	0.43	8.02	9.70	8.86

Avg: 0.39; Avg: 8.45.

Ventilator exhaust rate:	V = 1.5 m/s					
	PMV			PPD		
Testing sector:	Min	Max	Mean	Min	Max	Mean
Pole 1	0.32	0.48	0.40	7.10	9.80	8.45
Pole 2	0.03	0.47	0.25	5.00	9.60	7.30
Pole 3	0.29	0.46	0.38	6.70	9.80	8.20
Pole 4	0.31	0.46	0.39	7.00	9.60	8.30
Pole 5	0.31	0.47	0.39	6.90	9.70	8.30
Pole 6	0.36	0.48	0.42	7.80	9.70	8.75
Pole 7	0.36	0.48	0.42	7.70	9.60	8.65
Pole 8	0.38	0.48	0.43	7.90	9.60	8.75

Avg: 0.38; Avg: 8.34

Ventilator exhaust rate:	V = 2 m/s					
	PMV			PPD		
Testing sector:	Min	Max	Mean	Min	Max	Mean
Pole 1	0.30	0.47	0.38	6.80	9.50	8.15
Pole 2	0.01	0.45	0.23	5.00	9.40	7.20
Pole 3	0.28	0.45	0.37	6.60	9.40	8.00
Pole 4	0.29	0.44	0.36	6.70	9.30	8.00
Pole 5	0.28	0.45	0.37	6.70	9.50	8.10
Pole 6	0.35	0.45	0.40	7.50	9.50	8.50
Pole 7	0.34	0.45	0.39	7.40	9.30	8.35
Pole 8	0.35	0.45	0.40	7.50	9.40	8.45

Avg: 0.36; Avg: 8.09

Ventilator exhaust rate:	V = 3 m/s					
	PMV			PPD		
Testing sector:	Min	Max	Mean	Min	Max	Mean
Pole 1	0.28	0.46	0.37	6.62	9.20	7.91
Pole 2	-0.03	0.43	0.20	4.98	9.20	7.09
Pole 3	0.26	0.45	0.36	6.44	9.10	7.77
Pole 4	0.27	0.44	0.36	6.52	9.20	7.86
Pole 5	0.26	0.44	0.35	6.44	9.30	7.87
Pole 6	0.33	0.43	0.38	7.28	9.20	8.24
Pole 7	0.32	0.43	0.38	7.13	9.10	8.12
Pole 8	0.34	0.44	0.39	7.35	9.20	8.28

Avg: 0.36; Avg: 7.89

The present study defines the value for the metabolic or activity rate in present study was set at 1.2 met (70 W/m^2) and clothes thermal resistance was taken equal to 0.55 clo (0.085 W/m^2). Based on the data obtained from the various exhaust rate, the corresponding range of PMV and PPD vales are 0.39 - 0.35 and 8.45% - 7.89%, respectively. The calculated PMV-PPD index indicates that the indoor air quality is satisfactory for thermal comfort. The positive value of PMV shows the indoor environment is slightly warmer than the ideal condition but still within the acceptable range.

4.4. Improvement and Optimization

From the analysis it can be concluded that the installation of ventilator improves the indoor comfort level for all test cases. The standard room comfort requirement as per set out by ISO 7730 can be achieved. However the indoor temperature is still slightly warmer than the optimum condition. The result of this investigation also implies that to further enhance the room temperature for the optimum condition, higher exhaust rate are required. This can be done through a larger capacity ventilator or a hybrid roof top ventilator. Recent experiment investigation by Lai [40] reported a hybrid roof top ventilator that combines a small DC fan powered by PV cells can provide extra performance boost in the case of low wind speed but high sunlight.

5. CONCLUSION

The indoor air quality of a model house living room with installation of a rooftop turbine ventilator was simulated using CFD analysis. The results indicate that with the presence of the rooftop ventilator, the indoor air quality can meet the general requirements for comfort standards in various building codes. The results also showed that the comfort level increases with the exhaust rate induced by the turbine ventilator. This study presents a quantitative as well as qualitative study for indoor air quality that offered further scope to investigate the indoor air flow pattern inside a room. Thus a promising conclusion that can be drawn from this study is that CFD analysis could be used as a cost effective aid to future optimize the indoor air quality within a room.

REFERENCES

1 B. O. Brooks and W. F. Davis, "Understanding Indoor Air Quality," CRC Press Inc., Boca Raton, 1992.

2 G. Pillgram Larsen, "300000 pa Skolebenken for a Bedre Innemijoet," Statens Bygningstekniske Etat, Oslo, 1991.

3 O. Seppanen and J. Palonen, "The Effects of Indoor Climate on National Economy in Finland in Billoon FIM," FISIAW, SIY Rapportti 10, 1998.

4 A. Pissasale and N. A. Ahmed, "Theoretical Calibration of a Five Hole Probe for Highly Three Dimensional Flow," International Journal of Measurement Science and Technology, Vol. 13, 2002, pp. 1100-1107.

5 A. Pissasale and N. A. Ahmed, "Examining the Effect of Flow Reversal on Seven-Hole Probe Measurements," AIAA Journal, Vol. 41, No. 12, 2003, pp. 2460-2467.

6 A. Pissasale and N. A. Ahmed, "A Novel Method of Extending the Calibration Range of Five Hole Probe for Highly Three Dimensional Flows," Journal of Flow Measurement and Instrumentation, Vol. 13, No. 1-2, 2002, pp. 23-30.

7 A. Pissasale and N. A. Ahmed, "Development of a Functional Relationship between Port Pressures and Flow Properties for the Calibration and Application of MultiHole Probes to Highly Three-Dimensional Flows," Experiments in Fluids, Vol. 36, No. 3, March, 2004, pp. 422-436.

8 S. Shun and N. A. Ahmed, "Utilizing Wind and Solar Energy as Power Sources for a Hybrid Building Ventilation Device," Renewable Energy, Vol. 33, No. 6, 2008, pp. 1392-1397.

9 S. Shun and N. A. Ahmed, "Airfoil Separation Control Using Multiple Orifice Air Jet Vortex Generators," AIAA Journal of Aircraft, Vol. 48, No. 6, 2011, pp. 1994-2002.

10 N. Findanis and N. A. Ahmed, "The Interaction of an Asymmetrical Localised Synthetic Jet on a Side Supported Sphere," Journal of Fluids and Structures, Vol. 24, No. 7, 2008, pp. 1006-1020.

11 N. Findanis and N. A. Ahmed, "Three-Dimensional Flow Reversal and Wake Characterisation of a Sphere Modified with Active Flow Control Using Synthetic Jet," Advances and Applications in Fluid Mechanics, Vol. 9, No. 1, 2011, pp. 17-76.

12 N. A. Ahmed, "Wind Driven Natural-Solar/Electric Hybrid Ventilators," Wind Power, Section D: The Environmental Issues, In-Tech Organization, Austria, 2010.

13 J. Lien and N. A. Ahmed, "Wind Driven Ventilation for Enhanced Indoor Air Quality," Chemistry, Emission, Control, Radiaoactive Pollution and Indoor Air Quality, In-Tech Organization, Austria, 2011.

14 N. Findanis and N. A. Ahmed, "Wind Tunnel 'Concept of Proof' Investigations in the Development of Novel Fluid Mechanical Methodologies and Devices," Wind Tunnels and Experimental Fluid Dynamics Research, In-Tech Organization, Austria, 2011.

15 C. Wu and N. A. Ahmed, "Numerical Study of Transient Aircraft Cabin Flowfield with Unsteady Air Supply," AIAA Journal of Aircraft, Vol. 48, No. 6, 2011, pp. 1994- 2002.

16 N. A. Ahmed, R. L. Elder, C. P. Foster and J. D. C. Jones, "A Novel 3D Laser Anemometer for Boundary Layer Studies," 3rd International Symposium on Laser Anemometry, The Fluids Engineering Division, Vol. 55, pp. 175-178.

17 N. A. Ahmed, R. L. Elder, C. P. Foster and J. D. C. Jones, "Miniature Laser Anemometer for 3D Measurements," Journal of Measurement Science and Technology, Vol. 1, No. 3, 1990, pp. 272-276.

18 N. A. Ahmed and D. J. Wagner, "Vortex Shedding and Transition Frequencies Associated with Flow around a Circular Cylinder," AIAA Journal, Vol. 41, No. 3, 2003, pp. 542-544.

19 N. A. Ahmed, S. Hamid, R. L. Elder, C. P. Foster, J. D. C. Jones and R. Tatum, "Fibre Optic Laser Anemometry for Turbo Machinery Applications," Optics and Lasers in Engineering, Vol. 15, No. 2-3, 1992, pp. 193-205.

20 R. G. Simpson, N. A. Ahmed and R. D. Archer, "Near Field Study of Vortex Attenuation Using Wing Tip Blowing," The Aeronautical Journal, Vol. 102, 2002, pp. 117- 200.

21 R. G. Simpson, N. A. Ahmed and R. D. Archer, "Improvement of a Wing Performance Using Coanda Tip Jets," AIAA Journal of Aircraft, Vol. 37, No. 1, 2000, pp. 183-184.

22 A. Gatto, N. A. Ahmed and R. D. Archer, "Investigation of the Upstream End Effect of the Flow Characteristics of a Yawed Circular Cylinder," The RAeS Aeronautical Journal, Vol. 104, No. 1033, 2000, pp 253-256.

23 N. A. Ahmed and R. D. Archer, "Testing of a Highly Loaded Horizonatal Axis Wind Turbines Designed for Optimum Performance," International Journal of Renewable Energy, Vol. 25, No. 4, 2002, pp. 613-618.

24 N. A. Ahmed and R. D. Archer, "Performance Improvement of a Bi-Plane with Endplates," AIAA Journal of Aircraft, Vol. 38, No. 2, 2001, pp. 398-400.

25 N. A. Ahmed and J. Goonaratne, "Lift Augmentation of a Low Aspect Ratio Thick Wing at a Very Low Angle of Incidence Operating in Ground Effect," AIAA Journal of Aircraft, Vol. 39, No. 2, 2002, pp. 381-384.

26 J. Yen and N. A. Ahmed, "Parametric Study of Dynamic Stall Flow Field with Synthetic Jet Actuation," Journal of Fluids Engineering, Transactions of the ASME, Vol. 134, No. 7, 2012, Article ID: 071106.

27 Leadership in Energy & Environmental Design (LEED), Update December 2009.

28 S. J. Lien and N. A. Ahmed, "Effect of Inclined Roof on the Airflow Associated with a Wind Driven Turbine Ventilator," Energy and Building, Vol. 43, No. 2-3, 2011, pp. 358-365.

29 M. Bojic, F. Yik and T. Y. Lo, "Locating Air-Conditioners and Furniture inside Residential Flats to Obtain Good Thermal Comfort," Energy and Buildings, Vol. 34, No. 7, 2002, pp. 745-751.

30 W. N. Hien, W. Liping, A. N. Chandra, A. R. Pandey and X. L. Wei, "Effects of Double Glazed Facade on Energy Consumption, Thermal Comfort and Condensation for a Typical Office Building in Singapore," Energy and Buildings, Vol. 37, No. 6, 2005, pp. 563-572.

31 A. Stamou and I. Katsiris, "Verification and Application of a CFD Model for the Evaluation of Thermal Comfort in Office Spaces," Building and Environment, Vol. 41, No. 9, 2006, pp. 1171-1181.

32 T. Karimipanah, H. B. Awbi, M. Sandberg and C. Blomqvist, "Investigation of Air Quality, Comfort Parameters and Effectiveness for Two Floorlevel Air Supply Systems in Classrooms," Building and Environment, Vol. 42, No. 2, 2007, pp. 647-655.

33 S. J. Lien and N. A. Ahmed, "Numerical Simulation of Rooftop Ventilator Flow," Building and Environment, Vol. 45, No. 8, 2012, pp. 1808-1815.

34 Q. Chen, "Performance Prediction for Buildings: A Method Overview and Recent Applications," Building and Environment, Vol. 44, No. 4, 2009, pp. 848-858.

35 "PHOENICS 3.2 Software Package," CHAM, London, 1999.

36 "AEA Technology," CFX-5.5.1 User Guide, 8.19, Harwell, Didcot, Oxfordshire, OX110RA, 2002.

37 "Fluent User's Guide," Version 6.3, Fluent Inc., Lebanon, 2004.

38 Barbason and Reiter, "2010 ISO International Standard 7730. Moderate Thermal Environment: Determination of the PMV and PPD Indices and Specifications of the Condition for Thermal Comfort," International Organization for Standardization, Geneva, 2005.

39 ASHRAE, "ASHRAE Handbook Fundamentals," American Society of Heating Refrigerating and Air Conditioning engineers, Atlanta, 1997.

40 C. Lai, "Prototype Development of the Rooftop Turbine Ventilator Powered by Hybrid Wind and Photovoltaic Energy," Energy & Buildings, Vol. 38, No. 3, 2006, pp. 174-180.

CHAPTER 8

Indoor Air Quality in the United Arab Emirates

William E. Funk[1*], Joachim D. Pleil[2], Joseph A. Pedit[3], Maryanne G. Boundy[3], Karin B. Yeatts[4], David G. Nash[2], Chris B. Trent[5], Mohamed El Sadig[6], Christopher A. Davidson[3], David Leith[3]

[1]Department of Preventive Medicine, Feinberg School of Medicine, Northwestern University, Chicago, USA
[2]US Environmental Protection Agency, Research Triangle Park, USA
[3]Department of Environmental Sciences and Engineering, Gillings School of Global Public Health, University of North Carolina at Chapel Hill, Chapel Hill, USA
[4]Department of Epidemiology, Gillings School of Global Public Health, University of North Carolina at Chapel Hill, Chapel Hill, USA
[5]US Department of Housing and Urban Development, Washington DC, USA
[6]Department of Community Medicine, Faculty of Medicine and Health Sciences, United Arab Emirates University, Al Ain, UAE

ABSTRACT

Air quality was measured inside 628 United Arab Emirates (UAE) personal residences. Weekly average concentrations of carbon monoxide (CO), formaldehyde (HCHO), hydrogen sulfide (H_2S), nitrogen dioxide (NO_2), sulfur dioxide (SO_2), and three size fractions of particulate matter ($PM_{2.5}$, PM_c, and PM_{10}) were determined in each home. In a subset of the homes, measurements of outdoor air quality, ultrafine PM concentrations, and elemental PM concentrations were also made. Questionnaires were administered to obtain information on housing demographics and lifestyle habits. Air measurements were performed using simple and cost effective passive samplers. The 90th percentiles of indoor CO, HCHO, H_2S, NO_2, and SO_2 were 1.55 ppm, 0.05 ppm, 0.12 ppm, 0.01 ppm, and 0.05 ppm, respectively. Median indoor $PM_{2.5}$, PM_c, and PM_{10}, concentrations were 5.73 μg/m^3, 29.4 μg/m^3, and 35.2 μg/m^3, respectively. The median indoor concentration of ultrafine PM was 3.62×10^{10} particles/m^3. Indoor/outdoor ratios for PM were 0.44, 0.41, and 0.38 for ultrafine PM, $PM_{2.5}$, and PM_{10}, respectively. These values fall within the range of other indoor air studies findings conducted in developing countries. Air conditioning, smoking, and attached kitchens were significantly correlated with indoor levels of carbon monoxide. In addition, indoor concentrations of $PM_{2.5}$ and PM_{10} were significantly correlated with vehicles parked within five meters of the home, central air conditioning, and having attached kitchens. This is the first robust indoor air quality data set developed for the UAE. This study demonstrates that screening level tools are a good initial step for assessing air quality when logistical issues (distance, language, cultural, training) and intrusion into personal lives need to be minimized.

KEYWORDS

Indoor Air Quality, UAE, Particulate Matter, Pollution, Gases, Sources

1. INTRODUCTION

The discovery of vast oil resources in the UAE has enabled its citizens to progress within the past 50 years from a semi-nomadic existence in a harsh desert environment to a thriving lifestyle with vast, ultra-modern facilities and infrastructure. Although these economic and social changes were accompanied by great improvements in public health, concern exists that the rapid modernization may have created some detrimental environmental effects for the population. To study these concerns, the Environment Agency-Abu Dhabi commissioned a multidisciplinary, environmental health project that included epidemiologic [1] , nutritional [2] , and indoor air components. This third section, the environmental component of the study, supported the epidemiology study by characterizing indoor concentrations of gaseous and particulate pollutants in 628 Emirati residences.

Input from 56 environmental health stakeholders in the UAE prioritized environmental risks and interventions based on an expanded WHO burden of disease approach [3] . This effort revealed the leading concerns for mortality risks in the UAE were from exposures to ambient and indoor air pollution. Further, risk assessment analyses estimated that 290 deaths and more than 89,000 health care visits per year in the UAE may be related to exposure to poor indoor air quality. We note that these statistics are similar to those from other developed environments (e.g. New York, USA) and slightly lower than those found for London, UK [4] . However, as infectious diseases are reduced as important factors for human morbidity and mortality in developing countries, the effects on long-term latency diseases from environmental contamination associated with construction, production of goods, availability of private transportation, increases in consumption, etc. become more prominent concerns.

Previous studies in the US and Canada have found that indoor air is a concern because people spend more than 87% of their lives indoors [5] [6] . Emirati citizens may spend an even greater percentage of their time indoors because of high ambient temperatures and cultural factors that may limit outdoor activities. In addition, residential exposures may be of particular concern for vulnerable populations, such as developing fetuses [7] [8] , infants [9] , children [10] [11] , the elderly [12] , the and other susceptible individuals [13] .

Contaminants in the indoor air environment comprise a range of gases from combustion by-products to volatilized organic chemicals. Particulate matter arising from any combustion process[14] , or the ever-present desert sand may also pose potential impacts. This array of pollutants can arise from indoor activities or result from outdoor pollutants that penetrate into the indoor environment. The balance of indoor-outdoor contamination is a function of building tightness, construction materials, local roadway traffic, use and type of cooking appliances, cultural practices, lifestyle habits, and personal activities of the inhabitants.

Among the many health effects that have been linked to indoor air quality, asthma is perhaps the most significant. In the case of Emirati children, their increased prevalence of asthma during the past two decades is thought to be related to indoor air pollutants associated with changes in living habits, the presence of secondhand tobacco smoke in the residence [15] , urbanization [16] , and genetic susceptibility [17] - [19] . Therefore, an assessment of indoor air quality is essential to any overall evaluation of environmental health risks.

This study was designed to acquire representative measurements of indoor pollutants that could be used in association with on-going epidemiologic health evaluations of 628 UAE households. As a complement to one of the largest epidemiology studies in the Arabian Gulf region, a simple, inexpensive assessment of indoor air quality was required. For each home, the average indoor concentration over a one-week period was determined for CO, HCHO, H_2S, NO_2, SO_2, and three size fractions of particulate matter: $PM_{2.5}$, PM_c, and PM_{10}, which refer to particles sized less than 2.5 μm aerodynamic diameter (fine fraction), between 10 and 2.5 μm aerodynamic diameter (coarse fraction), and less than 10 μm aerodynamic diameter, respectively.

Additional particle samples were collected in 14 to 49 homes to evaluate indoor-outdoor pollutant ratios, ultrafine PM concentrations, and elemental PM composition. To complement the air quality measurements, questionnaires were administered in Arabic to selected family members to provide information on housing demographics and lifestyle habits. These data were used to assess health impacts associated with indoor air quality [1], and to pinpoint potential pollutant sources to inform future remediation efforts.

2. METHODS

2.1. Study Design

This study was reviewed and approved by the Institutional Review Boards of the University of North Carolina at Chapel Hill and the UAE University Faculty of Medicine. All informed consents and interviews with the study population were conducted by Arabic-speaking personnel using questionnaires written in Arabic.

The cross-sectional study design employed a nationally representative, stratified random sample of urban and rural Emirati residences from the seven Emirates (Abu Dhabi, Ajman, Dubai, Fujairah, Ras al-Khaimah, Sharjah, and Umm al-Quwain), as described by Yeatts et al. [1] . In brief, these 14 strata (7 Emirates × 2 locales) were divided into primary sampling units consisting of a census enumeration area in the urban areas or a village in rural settings. From the 2008 census data, the UAE Ministry of Economy randomly selected 120 primary sampling units across the country, with each unit containing at least eight Emirati households. Our study teams contacted 827 Emirati households; 628 households agreed to participate, yielding a response rate of 76%.

During the five-month period prior to the start of the study, all sampling equipment, protocols, and questionnaires were piloted in-country and modified as needed. Equipment and supplies were distributed to seven field sites. Field staff were recruited, trained, and tested in all aspects of the study and sampling protocols, including quality assurance and quality control procedures, the data management system, and cultural issues and sensitivities [20] .

Data collection required two home visits that occurred approximately seven days apart. At the first visit, following receipt of informed consent, the air monitoring equipment was deployed in a room where the family members spent a majority of their time together. One week later, the field staff returned to the residence to retrieve and enter data from the sampling equipment and to interview selected family members using a questionnaire format. Questions for the environmental exposure section were developed with guidance from the RIOPA survey [21] and the UAE Health and Lifestyle Survey [22] . Information was obtained on housing characteristics, residential history, potential indoor and outdoor environmental exposures, and behavioral factors such as smoking and incense use.

2.2. Adapting to Practical and Cultural Challenges

The planning and execution of indoor sample collection in this region required our awareness of several challenges unique to this study. First, we needed to accommodate the logistics and distances involved for supply/ resupply of sampling equipment, training of infield personnel, and execution of study objectives (deploying equipment, collecting samples, interpreting meta data, etc.). Secondly, we had to be respectful of cultural restrictions and religious conventions in dealing with the study participants within their own homes. Therefore, initial decisions were made to simplify deployed materials and equipment as much as possible. When studies are performed locally, for example, it is relatively easy to use experienced laboratory personnel as field operators, and to have a quick response for repair, replacement, and operation of complex instrumentation. However, for this study of hundreds of homes distributed across the UAE, the logistics required that in-country personnel assume primary roles of operating the study, and that the shipment of samples to US for analysis was minimized. As such, we opted to use a passive monitoring approach with diffusion tubes that could be read (color change) directly in the field for gas-phase species, and to use miniaturized passive diffusion/settling substrate assemblies that could be analyzed in the US for constituents and sizing of particulate matter.

Intrusions into the home life of study participants are always a concern, and perhaps a bit more complex in an Arab country where gender roles, personal activities (religious observations, smoking, incense use, etc.), athome chores (cooking, cleaning, etc.) are practiced differently than in the US. Therefore, the use of passive and non-intrusive sampling methods became an important feature to minimize such disturbances. Also, the use of long-term passive samplers allowed us to reduce the number of visits to the homes in contrast to active samplers where media are exchanged and flows calibrated typically every 24 hrs.

2.3. Indoor Air Pollutant Measurements—Gaseous Pollutants

A sampling strategy was developed and validated using colorimetric passive diffusion tubes (Gastec Corp., Kanagawa, Japan) to measure low indoor concentrations of CO, HCHO, H_2S, NO_2, and $SO_2$for one week [23] . This application contrasts to the conventional use of passive diffusion tubes where industrial concentrations of pollutants are evaluated during an eight-hour work shift.

Prior to use, extensive quality control was performed on random samples from different lots of the diffusion tubes to assess variance within and between lots. In the field, each diffusion tube served as its own blank; any tube that read non-zero prior to deployment was discarded. A second, randomly selected diffusion tube by gas type was deployed in each residence as a duplicate sample. At the conclusion of each seven-day period, all diffusion tubes were read inside the residence by two independent field staff members who compared the exposed tubes to a set of unopened tubes and to a set of laminated photographs that illustrated the proper color changes for each tube type. Results were compared between readers, and a single measurement for each tube type was recorded. Gaseous concentrations were then calculated using pollutant-specific algorithms [23] .

2.4. Indoor Air Pollutant Measurements—Particulate Matter

PM measurements were made using the UNC Passive Aerosol Sampler, which consists of for a scanning electron microscope (SEM) aluminum stub covered by a stainless steel mesh cap [24] . This lightweight (6.5 g)

PM sampler is approximately the size of a dime and collects particles by gravitational settling and diffusion. For automated SEM analysis, a polycarbonate substrate was applied to the SEM stub using Electrodag graphite paint; then the mesh cap was immediately placed on the stub to prevent contamination and secured with two small screws. The entire unit was placed in a plastic holder and covered with a clear plastic cap until deployment.

Sampling began with the removal of the plastic cap to expose the passive aerosol sampler to air. Similarly, sampling ended with the replacement of the protective cap. Following deployment, the exposed samplers and associated blanks were shipped to RJ Lee Group, Inc. (Monroeville, PA) for analysis of $PM_{2.5}$, PM_c, and PM_{10} using computer-controlled scanning electron microscopy (CCSEM), an automated imaging system that detects and counts particle with diameters greater than 0.1 μm. This information was used to determine the concentration and size distribution of PM in each sample.

Elemental analysis was conducted on passive PM samplers in 13 residences where indoor and outdoor samplers were deployed. For these samplers, energy dispersive spectroscopy was used to determine the elemental composition of the collected aerosol particles in each size fraction. Data were obtained for 19 elements: aluminum (Al), barium (Ba), calcium (Ca), carbon (C), chlorine (Cl), chromium (Cr), copper (Cu), iron (Fe), lead (Pb), magnesium (Mg), manganese (Mn), nickel (Ni), phosphorus (P), silicon (Si), sodium (Na), sulfur (S), titanium (Ti), and zinc (Zn).

Ultrafine PM was measured in 24 homes using the same passive aerosol sampler with a Pelco Formvar Carbon Type-B grid (Ted Pella, Inc., Redding, California) [25] . These ultrafine PM samples were analyzed at the UNC using field emission scanning electron microscopy operated at 125 kX. The analysis allowed particle number concentrations for ultrafine PM to be determined for each sample.

Quality assurance and quality control procedures were implemented for the PM aerosol samplers and followed throughout the study. In five percent of randomly selected homes, duplicate PM samplers were deployed to determine the precision of the method. Blanks were also deployed in five percent of randomly selected residences; protective caps were not removed from these samplers. Similarly, duplicate and blank ultrafine samplers were deployed in five percent of the residences. Chain of custody was established to document the distribution, retrieval, shipment, and analysis of all PM samplers.

2.5. Deployment of Samples

For the indoor environment, high-density polyethylene sampling blocks (9 cm × 30 cm × 2 cm) were designed and built to hold seven passive diffusion tubes and four passive aerosol samplers with their protective caps. To prevent tampering or injury from the glass diffusion tubes, each block was fitted with a stainless steel cage that was secured to a modified surveyor's tripod at a height of 1.3 meters above the ground. The tripod assembly was placed in a common room where family members spent the most time.

Outdoor passive PM samplers were deployed using a shelter of flat plates designed by Ott and Peters [26] to control air flow above the aerosol samplers and to protect them from weather. The flat plate shelter was adapted with extension screws to permit attachment of a circular polyethylene base (d = 10.2 cm) that held seven passive diffusion tubes. A 10 cm × 34 cm solar shield (Clear Dome Solar, San Diego, CA) was placed around the base to shield the diffusion tubes from direct sunlight. The shelter and base were also secured to a modified surveyor's tripod at a height of 1.3 meters above the ground and placed in a protected fenced-in area or courtyard. The tripod was weighted to prevent toppling.

All airborne pollutants were sampled for one week. For pollutant gases, sampling began by snapping off the top section of the passive diffusion tube at the designated scored mark. The sealed end of each glass tube was then inserted into its labeled position on the sampling block.

Similarly, for the passive aerosol sampler, removal of the protective cap and insertion of the sampler into position on the block or in the shelter constituted the start of the sampling period. Dates and times were recorded at the start and end of the sampling period.

2.6. Statistical Analyses

Indoor air pollution concentration means, medians, standard deviations, etc., were calculated using JMP software (JMP version 10, SAS Institute Inc., Cary, NC USA). Certain calculations of log-normal parameters were conducted using MS Excel (Excel: Mac 2011, Version 14.1.0, Redmond, WA, USA); some graphics were constructed using Graph Pad Prism (Prism 5: OS-X, 5.0c, La Jolla, CA, USA).

Associations between indoor air concentrations and environmental questionnaire data, such as second-hand smoke, air conditioning, etc., were examined using one sided t-tests. Because of the explorative nature of these analyses, corrections for multiple testing were not performed.

3. RESULTS

3.1. Duplicate and Blank Samples

Duplicate passive diffusion tubes that were deployed in 16% - 20% of homes for each gaseous pollutant had average relative standard deviations of 11%, 5%, 5%, 4%, and 8% for CO, HCHO, H_2S, NO_2, and SO_2, respectively.

For the PM passive samplers, no particles were detected on the blank samplers. The averages of the relative standard deviations for the 33 duplicate PM samples were 20%, 16%, and 15% for $PM_{2.5}$, PM_c, and PM_{10}, respectively. In examining the precision of these duplicate pairs, Arashiro and Leith (2013) found precision increased with higher PM concentrations due to better counting statistics. Although the passive sampler is not an EPA reference method, the precision of this device is close to the coefficient of variation of 10% set by the US EPA for operational precision.

3.2. Indoor Air Quality

Table 1 provides the median and 90th percentile indoor concentrations of the air pollutants. The table also shows the percentage of homes where the concentration of a pollutant was below the limit of quantification for the monitors used in this study. Because these measurements represent seven-day average concentrations made with screening-level devices, these measurements should be interpreted with caution when comparing results to international air quality standards based on short-term (1-hr), work-day (8-hr), or 24-hr timeframes. However, relative pollutant concentrations were used along with environmental questionnaire data to pinpoint potential sources of pollutants, and along with associated health questionnaires to evaluate potential health risks [1] . Ratios of indoor and outdoor concentrations were also examined for general trends, such as differences across rural versus urban locations.

Table 1 . Air pollutants measured indoors in the UAE.

Pollutant, units	Households Sampled	% Below LOQ	Median	90th Percentile
Ultrafine PM, particles/m^3	23	0	3.62×10^{10}	1.01×10^{11}
$PM_{2.5}$, μg/m^3	575	0	5.73	14.5
PM_c, μg/m^3	575	0	29.4	72.8
PM_{10}, μg/m^3	575	0	35.2	85.5
CO, ppm	625	0	0.77	1.55
HCHO, ppm	626	65	<0.01*	0.05
H_2S, ppm	626	83	<0.06*	0.12
NO_2, ppm	626	84	<0.01*	0.01
SO_2, ppm	625	62	<0.01*	0.05

*Approximate level of quantification.

3.3. Gaseous Pollutants

Table 1 shows that indoor concentrations of gaseous pollutants were mostly below the limit of quantification with the exception of CO. However, detectable concentrations of HCHO, H_2S, NO_2, and SO_2 were measured inside 35, 17, 16, and 38 percent of homes, respectively. Measurable levels of CO were present in all homes with a median indoor concentration of 0.76 ppm, and because CO is readily formed during indoor combustion processes (e.g., cooking, incense burning, and cigarette smoking), it was used as a general indicator for evaluating indoor combustion sources. Potential sources of CO from the environmental questionnaires show that split air conditioning, smoking, and attached kitchens were all significantly correlated with indoor levels of CO (Table 2). Split system air conditioning was found in 17% of the homes in our study and refers to a ductless system consisting of an outdoor condenser/compressor unit, an indoor wall-mounted unit with the air handler, and a remote control. Such a system allows cooling of individual rooms rather than an entire floor. Attached kitchens refer to the placement of the cooking area within the home, as is most common within the western world. Whereas the norm in the UAE was an "unattached kitchen" found in 69% homes, where food was prepared in a separate structure and brought to the residence at mealtimes. Outdoor construction within 100 meters of the home, vehicles parked within 5 meters of the home, open windows, portable heaters, use of natural gas in attached kitchens, and use of incense were not significantly associated with indoor concentrations of CO ($p > 0.05$).

Indoor and outdoor concentrations of gaseous pollutants measured in 49 homes showed that concentrations of CO were similar in both environments (Table 3). The median indoor/outdoor ratio was 0.95, which indicates that CO was higher outdoors than indoors. Higher concentrations of CO were also found in urban homes compared to rural residences for both indoor and outdoor samples (Table 4).

Table 2. Questionnaire variables as predictors of indoor air concentrations of CO, $PM_{2.5}$, and PM_{10}.

Questionnaire Variables	Expected value CO (ppm)			Expected value $PM_{2.5}$ ($\mu g/m^3$)			Expected value PM_{10} ($\mu g/m^3$)		
	no	yes	p	no	yes	p	no	yes	p
Motorized vehicles routinely park within 5 meters of home	0.954 (218)	0.919 (362)	0.803	7.21 (206)	7.48 (329)	**0.030**	45.6 (206)	46.1 (329)	**0.038**
Split system air conditioning units, not vented to outside	0.900 (504)	1.145 (72)	**0.001**	7.22 (464)	8.41 (67)	0.032	45.5 (464)	48.0 (67)	0.105
Central air conditioning unit	0.928 (461)	0.936 (115)	0.631	8.12 (423)	4.38 (108)	**1.000**	50.9 (423)	24.3 (108)	**1.000**
Kitchen attached to main living areas	0.879 (363)	1.056 (156)	**0.005**	8.11 (327)	6.24 (149)	**1.000**	51.4 (327)	38.4 (149)	**1.000**
Windows opened daily	0.969 (196)	0.914 (382)	0.884	7.38 (181)	7.43 (353)	0.189	46.6 (181)	45.8 (353)	0.353
Construction within 100 meters of home	0.907 (314)	0.965 (260)	0.188	7.51 (282)	7.28 (248)	0.659	45.8 (282)	46.0 (248)	0.451
Tobacco products smoked inside home	0.881 (353)	1.066 (117)	**0.005**	7.42 (322)	7.73 (110)	0.338	44.8 (322)	50.9 (110)	0.101
Incense used once or more per week	0.960 (36)	0.928 (483)	0.759	8.80 (34)	7.44 (442)	0.683	49.1 (34)	47.3 (442)	0.233
Portable electric heater used	0.929 (541)	0.972 (34)	0.457	7.39 (496)	6.76 (34)	0.550	45.3 (496)	49.2 (34)	0.058
Natural gas cooking equipment with kitchen attached to main living areas	0.894 (20)	1.081 (136)	0.117	7.70 (20)	6.05 (129)	0.669	40.6 (20)	38.2 (129)	0.292

*One-sided t-tests comparing air concentrations with and without the presence of questionnaire variables. p-values less than 0.05 confirms that "yes" is significantly greater than "no" responses. p-values greater than 0.95 confirms that "yes" is significantly less than "no" responses.

Table 3. Air pollutants measured indoors and outdoors in paired homes.

Pollutant, units	Households Sampled	Indoor Median	Indoor 90th Percentile	Outdoor Median	Outdoor 90th Percentile
Ultrafine PM, part./m^3	14	2.9×10^{10}	1.2×10^{11}	6.6×10^{10}	1.4×10^{11}
$PM_{2.5}$, $\mu g/m^3$	39	6.34	13.3	15.4	36.3
PM_c, $\mu g/m^3$	39	31.1	72.3	79.7	178.6
PM_{10}, $\mu g/m^3$	39	36.6	83.0	96.8	213.7
CO, ppm	49	0.73	1.25	0.77	1.29
HCHO, ppm	49	<0.01*	0.05	<0.01*	0.06
H_2S, ppm	49	<0.06*	0.12	<0.06*	0.14
NO_2, ppm	49	<0.01*	0.01	<0.01*	0.01
SO_2, ppm	49	<0.01*	0.05	0.01	0.06

*Approximate level of quantification.

Table 4. Air pollutants measured indoors in rural and urban locations in the UAE.

Pollutant, units	Rural, n	Rural Median	Rural 90th Percentile	Urban, n	Urban Median	Urban 90th Percentile
$PM_{2.5}$, µg/m³	245	6.06	17.9	330	6.0	13.5
PM_c, µg/m³	245	32.9	86.9	330	31.4	69.0
PM_{10}, µg/m³	245	41.3	106	330	38.5	83.5
CO, ppm	269	0.75	1.31	356	0.77	1.78
HCHO, ppm	269	<0.01*	0.05	357	<0.01*	1.03
H_2S, ppm	269	<0.06*	0.12	357	<0.06*	0.10
NO_2, ppm	269	<0.01*	0.01	357	<0.01*	0.01
SO_2, ppm	268	<0.01*	0.05	357	0.01	0.04

*Approximate level of quantification.

3.4. Particulate Matter

In the 39 homes where PM was measured both indoors and outdoors the median indoor concentrations of $PM_{2.5}$, PM_c, and PM_{10} were 6.34 µg/m³, 31.1 µg/m³, and 36.6 µg/m³, respectively (Table 3).

Indoor concentrations of $PM_{2.5}$ and PM_{10} were highly correlated, with a Pearson R = 0.84. The median outdoor concentrations of $PM_{2.5}$, PM_c, and PM_{10} were 15.4 µg/m³, 79.7 µg/m³, and 96.8 µg/m³, respectively. Concentrations of particles were lower indoors than outdoors; the ratios of the indoor to outdoor median $PM_{2.5}$, PM_c, and PM_{10} concentrations in these 39 paired samples were 0.41, 0.39, and 0.38, respectively. Indoor PM concentrations were higher in rural areas than urban locations, but outdoor PM concentration were not significantly different between rural and urban locations (Table 4).

Potential indoor sources of PM were investigated using environmental questionnaires to assess associations between indoor PM concentrations and questionnaire variables (Table 2). Indoor concentrations of $PM_{2.5}$ and PM_{10} were significantly associated with vehicles being parked within five meters of the home, use of central air conditioning, and with kitchens that were located within the residence.

Opening of windows, construction within one hundred meters of the house, cooking fuel type, cigarette smoking, and use of incense were not significantly correlated with indoor PM ($p > 0.05$). However, this lack of association may be due in part to the polycarbonate substrate used on the UNC Passive Aerosol Samplers, which can underestimate combustion-related (carbonbased) PM.Table 3 presents the 14 residences with paired indoor and outdoor ultrafine PM measurements. Concentrations are reported as the number of particles per unit volume (particles/m^3), which is the most standard convention for reporting ultrafine PM. The median concentrations of ultrafine PM in the indoor and outdoor samples were 2.9×10^{10} particles/m^3 and 6.6×10^{10} particles/m^3, respectively. In the 14 homes with paired indoor and outdoor measurements, the median indoor/outdoor ratio of ultrafine PM was 0.44.

3.5. Elemental Composition

Elemental analysis of the PM samples in Figure 1 shows a higher percentage of carbon was observed indoors, which may reflect indoor combustion sources such as smoking or burning incense. In contrast, the percentage of silicon was higher outdoors, which may reflect a higher prevalence of sand-based particles. Overall, differences for carbon and silicon were small, and indoor and outdoor profiles of the seven most abundant elements were similar, which is suggestive of a common PM source. This finding may be the result of infiltration of outdoor PM into the indoor environment. In addition, Figure 2 shows that chromium, nickel, copper, and zinc concentrations were higher indoors than outdoors, while lead was higher in the outdoor samples. These findings suggest that while infiltration of outdoor air may be a major source of indoor PM, localized sources of PM indoors and outdoors likely exist and provide differing trace-element profiles.

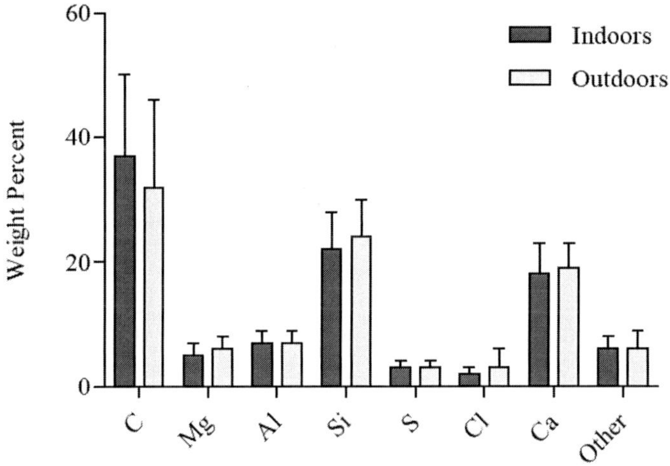

Figure 1. Weight percentages of the seven most abundant elements measured indoors and outdoors in 13 homes in the UAE. Error bars represent the standard deviations of the means.

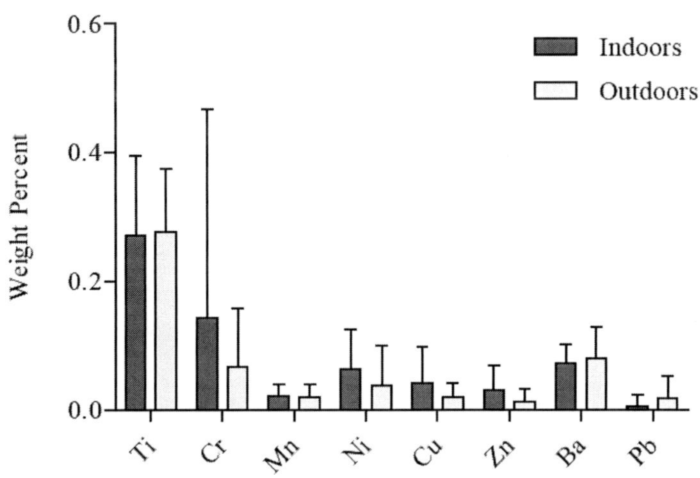

Figure 2. Weight percentages of eight less abundant elements measured indoors and outdoors in 13 paired homes in the UAE. Error bars represent the standard deviations of the means.

4. DISCUSSION

These results point out the complexity of indoor exposure patterns. While it is not possible to state unequivocally and clearly the causes of all indoor concentrations reported here, this discussion will focus on what can be said, and present a comparison with results of studies of indoor air quality in other parts of the world to place these finding in perspective.

This indoor air quality assessment in the UAE is the largest of its kind in the Middle East region. We evaluated 628 homes within the UAE, successfully collecting hundreds of measurements for five gaseous pollutants (CO, HCHO, H_2S, NO_2, and SO_2) and four size fractions of PM (Ultrafine PM, $PM_{2.5}$, PMc, and PM_{10}). Our targeted pollutants were selected based on common exposures that occupants in the UAE might encounter from sources such as infiltration of ambient air pollution from nearby industrial activity and traffic, as well as key indoor sources such as tobacco smoking and incense burning.

Indoor Particulate Matter: Detailed Analysis

In Table 5 we compare indoor concentrations of $PM_{2.5}$ measured in the UAE with other studies throughout different parts of the world. While these comparisons are meant to put the results of this study in context, we note that measurements made in this study were performed with screening-level samplers and that rigorous comparisons with other studies should be done with caution.

One recent study performed by the US EPA of the urban environment of Detroit, Michigan—the Detroit Exposure and Aerosols Research Study (DEARS)—spanned 3 winter and 3 summer seasons from 2004 to 2007 (Williams, 2008). Figure 3(a) shows the frequency distribution of the 628 indoor, week-long $PM_{2.5}$ measurements made in the UAE regardless of urban vs. rural locations and other meta-data. The figure shows that the data appear lognormal (which is expected for environmental data) and that the geometric mean (GM) and geometric standard deviation (GSD) are 5.79 μg/m^3 and 2.33 μg/m^3, respectively. Figure 3(b) shows the analogous $PM_{2.5}$ data from 539 measurements made in 134 urban Detroit homes from the DEARS study selected to represent the summer months of 2004-2006. Figure 3(b) shows the data are also lognormal in character but shifted somewhat higher with geometric mean (GM) and geometric standard deviation (GSD) of 16.24 μg/m^3 and 1.88 μg/m^3, respectively. Although the UAE homes experienced approximately the same outdoor $PM_{2.5}$ levels as Detroit, (UAE: GM = 15.4 μg/m^3, DEARS: GM = 14.0 μg/m^3), their indoor levels were a factor of 2.5 lower (GM = 5.79 μg/m^3 vs. 16.2 μg/m^3). Detroit is located in the northern part of the midwestern US where most homes were not air conditioned (cooled) in the summer; occupants generally used passive ventilation in the form of fans and open windows, which results in similar indoor and outdoor $PM_{2.5}$ concentrations. The UAE homes may have been more sealed due to the high ambient temperatures and the ever-present desert climate. UAE homes also tend to have separated kitchens, which further reduces sources of indoor $PM_{2.5}$.

Table 5. Comparison of $PM_{2.5}$ concentrations in studies conducted in different parts of the world.

Indoor Air Study	Mean, μg/m^3	CV	GM or median	GSD	Upper 95%	Reference
UAE	7.7	**1.2**	**5.8**	**2.3**	**23.3**	
USA, MI (DEARS)	18.7	1.1	13.8	2.2	49.3	Williams *et al*., 2009
USA, CA (RIOPA)	16.2	0.6	14.5	1.7	35.3	Meng *et al*., 2005
USA, TX (RIOPA)	17.1	0.7	13.4	1.9	40.0	Meng *et al*., 2005
USA, NJ (RIOPA)	20.1	0.8	15.7	2.0	48.5	Meng, *et al*., 2005
Kuwait	41.9	0.5	31.5	1.6	68.4	Yassin *et al*., 2012
Alexandria, Egypt	53.5	0.3	49.5	1.3	78.4	Abdel-Salam *et al*., 2013
Hong Kong, China	58.4	0.3	56.7	1.3	87.4	Cao *et al*., 2005
Bangkok, Thailand	137.0	0.2	131.2	1.2	184.9	Tsai *et al*., 2000
Haryana, India	718.0	0.5	568.0	1.6	1262.6	Mukhopadhay *et al*., 2012
Pakistan	2740.0	0.8	1750.5	2.0	5282.3	Siddiqui *et al*., 2009

Note: The values in this table were estimated from the respective published data summaries and harmonized under the assumption that measurements were log-normally distributed.

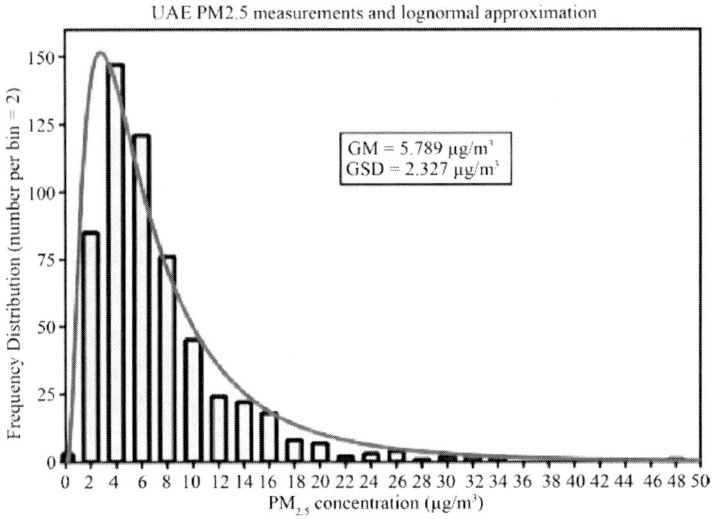

(a)

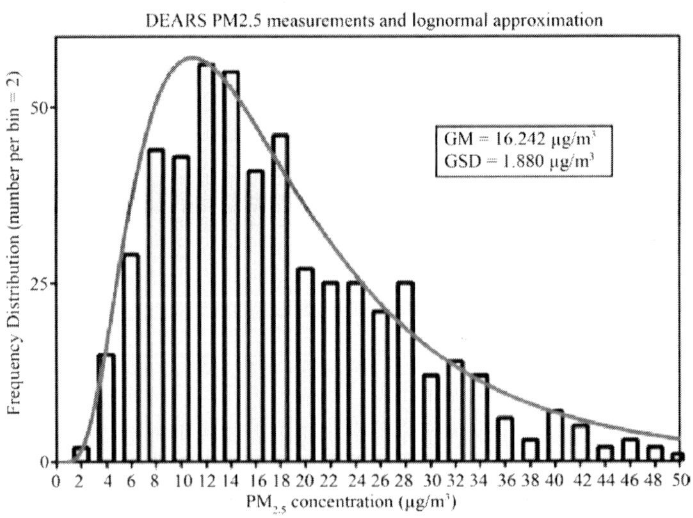

(b)

Figure 3. (a)/(b) Detailed comparison of distributions and levels of indoor air measurements of $PM_{2.5}$ in UAE and DEARS residential exposure studies. Both data sets show expected log-normal character; the UAE results are about 2.5 times lower most likely due to differences in ventilation (sealed AC vs. open windows) and differences in cooking (detached vs. attached kitchens).

Indoor air concentrations of CO ranged from 0.13 - 5.8 ppm, with a median value of 0.77 ppm. In homes where both indoor and outdoor measurement were made, the I/O ratio was 0.95, indicating that levels were slightly higher outdoors. This ratio was higher than the I/O ratio observed in a study performed in four sites in Saudi Arabia, which reported an average I/O ratio of 0.63 [27] . In this study, Rowe et al. reported automobile traffic as the major sources of outdoor CO, and cooking with gas stoves, use of kerosene heaters, charcoal burners, waterpipes, and cigarette smoking as the major sources for indoor CO [27] . Similarly, we found smoking and attached kitchens were significantly correlated with indoor levels of CO (Table 2).

We collected paired indoor and outdoor measurements for $PM_{2.5}$, PM_C, and PM_{10} at 39 homes and calculated an indoor-outdoor ratio (I/O) as an indicator of which environment may be the stronger contributor of PM. In our study, the median I/O ratios were 0.41, 0.39 and 0.38 for $PM_{2.5}$, PM_C, and PM_{10}, respectively. Rural areas had slightly higher median concentrations and wider ranges of concentrations in the upper quartile than urban locations. Potential determinants for this finding may include less development in rural areas to shelter homes from blowing sand and dust, older construction that allows for greater infiltration of ambient air pollution, or possibly different occupant activities and behaviors than in urban locations. Our I/O ratios indicate that generally outdoor PM concentrations exceeded those indoors, although this finding does not negate the contribution of indoor sources of exposures to the health burden of UAE residents.

In the RIOPA study [21] of 306 homes located in Los Angeles, CA; Elizabeth, NJ; and Houston, TX, the authors reported a median I/O for $PM_{2.5}$ of 0.70, with an interquartile range of 0.54 - 0.87. The RIOPA I/O results are reasonably comparable to those that we obtained from our study; however, it should be noted that the RIOPA study collected their samples continuously over a 48-hr period, while our samples were collected passively and integrated over a 7-day averaging period. The longer averaging time of our samples may tend to smooth out any episodic peak concentrations.

Geller et al. [28] calculated I/O ratios for 13 homes in the southern California desert region during winter and spring when occupants would be more likely to open windows and rely on natural ventilation. Measurements were collected in each home over 3 - 4 consecutive 23-hour periods. The topography and climate trends in this southern California region are comparable to those encountered in the UAE, since our study occurred during late fall, winter, and spring when temperatures in the UAE were more moderate. Geller and colleagues reported median I/O ratios of 0.92 and 0.54 for $PM_{2.5}$ and PM_{10}, respectively [28] . The median indoor $PM_{2.5}$ concentration reported by the authors was 4.6 μg/m^3 with a range of 0.6 - 20.2 μg/m^3, and the median indoor PM_{10} concentration was 13.5 μg/m^3 with a range of 4.2 - 49.3 μg/m^3.

We collected paired indoor-outdoor ultrafine PM samples for 14 homes and calculated an I/O ratio of 0.44. Ultrafine particles have a greater propensity to deposit deeper into the lung due to their small size and greater diffusion coefficient, which makes characterizing their contribution to indoor environments critical to understanding indoor air exposures. Ultrafine PM is associated with several ambient and indoor sources, including tobacco smoke and incense, both of which were commonly reported exposure sources in our study. For example, thirty-five percent of men surveyed in our study identified themselves as current smokers, and secondhand smoke exposures were reported by 19% of the heads of households who were surveyed [1] . Furthermore, almost half of the households reported daily incense use and over 80% burned incense in the home at least once a week [1] .

The overall median ultrafine PM number concentration from our study was 3.6×10^{10} particles/m^3. As a comparison, Nazaroff et al. [29] measured ultrafine PM concentration in seven homes in the San Francisco region and reported concentrations ranging from 3.7×10^9 particles/m^3 to 2.8×10^{10} particles/m^3. The authors attributed the particles to both ambient and indoor sources, including cooking and furnace use. Afshari et al. [30] conducted a chamber study for ultrafine particles associated with several different activities, including cooking, smoking cigarettes, and candle burning. The authors reported the highest particle number concentration of 2.41×10^{11}particles/m^3 when burning a beeswax candle in the chamber. Cigarette smoke resulted in a slightly lower particle concentration of 2.18×10^{11} particles/m^3. While we did not find a significant association between PM and waterpipe smoking, indoor air concentrations of ultrafine PM may have been underestimated due to the type of samplers used in our study and warrants further investigation. Waterpipe, or hookah, smoking is gaining popularity around the world and maintains a prominent role in Arabic culture, particularly among younger women who view it as having a less negative stigma than cigarette smoking [31] . Daher et al. [32] stated that pollutant levels in the sidestream (i.e., secondhand) smoke generated from one hookah session lasting one-hour were comparable to the smoking sessions of 2 - 5 cigarette smokers. The authors estimated that four repeated waterpipe sessions resulted in 3.99×10^{12} particles/waterpipe.

Incense use is another activity firmly grounded in the UAE and Arabic culture that may present exposures of concern to ultrafine PM. An experimental study by See et al. [33] focused on the physical characteristics of four types of incense commonly used by Taoists, Buddhists, Hindus, and Shinto followers. The authors calculated particle number concentrations ranging from 1.1×10^{12}particles/m^3 to 2.4×10^{12} particles/m^3 during incense burning. Our indoor ultrafine PM data reported in Table 1 and Table 3 are within the range of the study data of other researchers, and the activities that are reported to have occurred in our study homes are consistent with those evaluated in these other studies. Nevertheless, the ambient concentrations of ultrafine PM were somewhat elevated in our study, highlighting the importance of additional research in the UAE on this topic to accurately apportion these exposures.

In 14 homes in the UAE, indoor and outdoor elemental composition was examined. The composition profiles for indoor and outdoor samples were similar, which suggests that ambient infiltration may play a role in indoor PM concentrations. Carbon and silicon were identified as the most abundant elements. The presence of carbon inside the home is indicative of a combustion source, possibly from indoor sources such as tobacco smoke or incense. Cooking with natural gas may also be a source for some homes, but it may not be as prominent a source as one would find in the US or Europe because the kitchens of most Emirati homes are located in a separate building (e.g. 69% of homes reported having detached kitchens in this study). Ambient sources of carbon, including automobile exhaust from traffic, may also infiltrate into homes. The higher indoor presence of carbon in urban compared with rural areas may be due to the higher density traffic and lends support to the suggestion of ambient infiltration. Conversely, silicon, which is the primary element in sand, was more abundant in rural PM samples, possibly due to less dense neighborhoods and smaller buildings that permit sand to blow more freely and infiltrate into homes. However, as noted previously, the total concentration of carbon may have been underestimated due to the polycarbonate-coated SEM stubs used in this study, which are less sensitive to carbonaceous particles. This sample media bias may explain the lack of association found between activities such as smoking and burning incense indoors with indoor concentrations of PM.

Several trace metals were also measured including Ti, Cr, Mn, Ni, Cu, Zn, Ba, and Pb. Levels of Cr, Ni, Cu, and Zn were higher for indoor samples than outdoors. Sources of these metals could be of ambient or indoor origin, but additional research is needed to more definitively answer this question. Possible indoor sources include tobacco smoke, incense fumes, or burning candles. Chromium, Ni, Cu and Zn have been identified in cigarette smoke [34]. A recent study in Kuwait also examined the presence of trace metals in indoor air [35] . Although different elements were more prominent in that study, the investigators reported a similar trend to ours with trace metals being generally more abundant in indoor residential environments.

The pollutant concentration patterns reported here are clearly complex, so that definitive, general statements about causative factors cannot be made. What does emerge from this study is an understanding that: (1) concentrations of the indoor pollutant gases measured here were usually, but not always, relatively low (Table 1), (2) relationships exist between indoor concentrations of both gases and PM with lifestyle parameters such as smoking, the type of air conditioning used, and whether the residence has an attached or unattached kitchen (Table 2), (3) concentrations of PM, particularly PM_{10}, could sometimes be relatively high (Table 1), and (4) that one important source of indoor PM appears to be outdoor PM (Figure 1).

5. CONCLUSION

Our study successfully measured gaseous and PM air concentrations in 628 homes in the UAE as a first step to characterize indoor air quality in this widely understudied region. We found that overall, indoor air quality in UAE residences is quite similar to other developed nations; there were no remarkable outliers other than the slightly higher coarse particulate fraction (PM_c), which is likely attributable to the preponderance of sand in the desert environment. The passive sampling strategy developed for this study was minimally-invasive and inexpensive, yet provided reasonable data in comparison with other studies, such as the DEARS by the US EPA in Detroit, MI. This approach will help to facilitate the acceptance and feasibility of additional population-based studies of air quality in other international scenarios where logistics are more challenging. The wealth of data collected during this campaign can be used to develop further research studies to answer questions that have been spawned by these results, including apportionment of exposures, quantifying the contribution of ambient infiltration to indoor pollutant concentrations, levels of personal exposures, and understanding the effect of household activities on their exposures. While the air measurements reported in this study should be interpreted with caution given the screening-level precision of the samplers used, these relative estimations of air quality can successfully be used to evaluate the effects of indoor air pollution on human health [1] , and are a valuable resource for identifying potential sources of airborne pollutants in population-based research.

ACKNOWLEDGEMENTS

We thank our many collaborators in the UAE who contributed to this study, and the participants who graciously supported this work. We are grateful to the Collaborative Studies Coordinating Center at the University of North Carolina for their statistical assistance. This study was supported by funds from the Environment Agency-Abu Dhabi. This work has been subjected to U.S. Environmental Protection Agency review and approved for publication.

REFERENCES

1 Yeatts, K.B., et al. (2012) Indoor Air Pollutants and Health in the United Arab Emirates. Environmental Health Perspectives, 120, 687-694.

2 Ng, S.W., et al. (2011) Nutrition Transition in the United Arab Emirates. European Journal of Clinical Nutrition, 65, 1328-1337.

3 Willis, H.H., et al. (2010) Prioritizing Environmental Health Risks in the UAE. Risk Analysis: An Official Publication of the Society for Risk Analysis, 30, 1842-1856.

4 Gibson, J.M. and Farah, Z.S. (2012) Environmental Risks to Public Health in the United Arab Emirates: A Quantitative Assessment and Strategic Plan. Environmental Health Perspectives, 120, 681-686.

5 Klepeis, N.E., et al. (2001) The National Human Activity Pattern Survey (NHAPS): A Resource for Assessing Exposure to Environmental Pollutants. Journal of Exposure Analysis and Environmental Epidemiology, 11, 231-252.

6 Leech, J.A., et al. (2002) It's about Time: A Comparison of Canadian and American Time-Activity Patterns. Journal of Exposure Analysis and Environmental Epidemiology, 12, 427-432.

7 Choi, H., et al. (2012) Fetal Window of Vulnerability to Airborne Polycyclic Aromatic Hydrocarbons on Proportional Intrauterine Growth Restriction. PloS One, 7, Article ID: e35464.

8 Perera, F.P., et al. (2011) Polycyclic Aromatic Hydrocarbons-Aromatic DNA Adducts in Cord Blood and Behavior Scores in New York City Children. Environmental Health Perspectives, 119, 1176-1181.

9 Pickett, A.R. and Bell, M.L. (2011) Assessment of Indoor Air Pollution in Homes with Infants. International Journal of Environmental Research and Public Health, 8, 4502-4520.

10 Neas, L.M., et al. (1991) Association of Indoor Nitrogen Dioxide with Respiratory Symptoms and Pulmonary Function in Children. American Journal of Epidemiology, 134, 204-219.

11 Heinrich, J. (2010) Air Pollution, Asthma and Allergies in Children. Occupational and Environmental Medicine, 67, 290-291.

12 Simoni, M., et al. (2003) Indoor Air Pollution and Respiratory Health in the Elderly. European Respiratory Journal, 21, 15S-20S.

13 Bernstein, J.A., et al. (2008) The Health Effects of Nonindustrial Indoor Air Pollution. Journal of Allergy and Clinical Immunology, 121, 585-591.

14 Polidori, A., et al. (2006) Fine Organic Particulate Matter Dominates Indoor-Generated PM2.5 in RIOPA Homes. Journal of Exposure Science & Environmental Epidemiology, 16, 321-331.

15 Hassan Mahboub, B.H., et al. (2010) Asthma Insights and Reality in the United Arab Emirates. Annals of Thoracic Medicine, 5, 217-221.

16 Mahboub, B.H., Al-Hammadi, S., Rafique, M., Sulaiman, N., Pawankar, R., Al Redha, A.I. and Mehta, A.C. (2012) Population Prevalence of Asthma and Its Determinants Based on European Community Respiratory Health Survey in the United Arab Emirates. BMC Pulmonary Medicine, 12, 4.

17 Joseph, M., Zoubeidi, T., Al-Dhaheri, S.M., Al-Dhaheri, A.A., Al-Dhaheri, A.A., Al-Kaabi, F.M., Al-Muhairi, S.J. and Joseph, J. (2009) Paternal Asthma Is a Predictor for Childhood Asthma in the Consanguineous Families from the United Arab Emirates. Journal of Asthma, 46, 175-178.

18 Alsowaidi, S., Abdulle, A. and Bernsen, R. (2010) Prevalence and Risk Factors of Asthma among Adolescents and Their Parents in Al-Ain (United Arab Emirates). Respiration, 79, 105-111.

19 Alsowaidi, S., Abdulle, A., Bernsen, R. and Zuberbier, T. (2010) Allergic Rhinitis and Asthma: A Large Cross-Sectional Study in the United Arab Emirates. International Archives of Allergy and Immunology, 153, 274-279.

20 Yeatts, K.B., et al. (2012) Conducting Environmental Health Research in the Arabian Middle East: Lessons Learned and Opportunities. Environmental Health Perspectives, 120, 632-636.

21 Weisel, C.P., et al. (2005) Relationships of Indoor, Outdoor, and Personal Air (RIOPA). Part I. Collection Methods and Descriptive Analyses. Research Report, 130, 1-107; discussion 109-127.

22 Badrinath, P., Ghazal-Aswad, S., Parfitt, D. and Osman, N. (2004) Cultural and Ethnic Barriers in Conducting Research. Factors Influencing Menarche in the United Arab Emirates. Saudi Medical Journal, 25, 1626-1630.

23 Nash, D.G. and Leith, D. (2010) Use of Passive Diffusion Tubes to Monitor Air Pollutants. Journal of the Air & Waste Management Association, 60, 204-209.

24 Wagner, J. and Leith, D. (2001) Passive Aerosol Sampler. Part 1: Principle of Operation. Aerosol Sciences and Technology, 34, 186-192.

25 Nash, D.G. and Leith, D. (2010) Ultrafine Particle Sampling with the UNC Passive Aerosol Sampler. Aerosol Science and Technology, 44, 1059-1064.

26 Ott, D.K. and Peters, T.M. (2008) A Shelter to Protect a Passive Sampler for Coarse Particulate Matter, PM10-2.5. Aerosol Science and Technology, 42, 299-309.

27 Rowe, D.R., Aldhowalia, K.H. and Mansour, M.E. (1989) Indoor-Outdoor Carbon-Monoxide Concentrations at 4 Sites in Riyadh, Saudi-Arabia. JAPCA-The Journal of the Air & Waste Management Association, 39, 1100-1102.

28 Geller, M.D., Chang, M., Sioutas, C., Ostro, B.D. and Lipsett, M.J. (2002) Indoor/Outdoor Relationship and Chemical Composition of Fine and Coarse Particles in the Southern California Deserts. Atmospheric Environment, 36, 1099-1110.

29 Nazaroff, W.W., Bhangar, S. and Mullen, N. (2010) Ultrafine Particle Concentrations and Exposures in Northern California Houses. Abstracts of Papers of the American Chemical Society, Washington DC, 239.

30 Afshari, A., Matson, U. and Ekberg, L.E. (2005) Characterization of Indoor Sources of Fine and Ultrafine Particles: A Study Conducted in a Full-Scale Chamber. Indoor Air, 15, 141-150.

31 Nakkash, R.T., Khalil, J. and Afifi, R.A. (2011) The Rise in Narghile (Shisha, Hookah) Waterpipe Tobacco Smoking: A Qualitative Study of Perceptions of Smokers and Non Smokers. BMC Public Health, 11, 315.

32 Daher, N., Saleh, R., Jaroudi, E., Sheheitli, H., Badr, T., Sepetdjian, E., Al Rashidi, M., Saliba, N. and Shihadeh, A. (2010) Comparison of Carcinogen, Carbon Monoxide, and Ultrafine Particle Emissions from Narghile Waterpipe and Cigarette Smoking: Sidestream Smoke Measurements and Assessment of Second-Hand Smoke Emission Factors. Atmospheric Environment, 44, 8-14.

33 See, S.W., Balasubramanian, R. and Joshi, U.M. (2007) Physical Characteristics of Nanoparticles Emitted from Incense Smoke. Science and Technology of Advanced Materials, 8, 25-32.

34 Bernhard, D., Rossmann, A. and Wick, G. (2005) Metals in Cigarette Smoke. IUBMB Life, 57, 805-809.

35 Bu-Olayan, A. and Thomas, B. (2010) Validating Trace Metals Levels in PM2.5 Aerosols from Indoor and Outdoor Polluting Premises. American Journal of Environmental Sciences, 6, 224-229.

CHAPTER 9

Wind Driven Ventilation for Enhanced Indoor Air Quality

Jason Lien[1] and N.A. Ahmed

[1] School of Mechanical and Manufacturing Engineering University of New South Wales, Kensington, NSW 2052 , Australia

1. INTRODUCTION

The last century has seen tremendous progress in technological development that is underpinning the progress of the modern day human civilization. Unfortunately, however, such progress is giving rise to unwanted problems of grave consequences that have the potential to destroy the very environment which sustains life. There is now a real eagerness and greater awareness in the general public to look for alternative natural energy systems and products to help alter the present energy use patterns that have lead to this dilemma.

United Nations' conferences in Rio in 1992 and Johannesburg in 2002 have brought the issue of environment sustainability to the forefront in international circles. Subsequently the organization has defined goals that buildings should seek to achieve in order to obtain green building recognition that include goals such as increase in reliability, increase in indoor air quality, decrease in natural resource use, considerable decrease of energy costs over the lifetime of the building, improving comfort due to improved energy efficiency in building and raise of employment as a result of increased activity in energy improvements in buildings. These benefits will theoretically take care of any type of increase (typically 3-5%) in construction costs and making improvements will have a direct positive impact upon life-cycle costs.

The present chapter is an attempt to advance the cause of sustainable living and improve the quality and comfort of human existence through the greater use of environment friendly wind driven ventilation. Most of the materials presented are based on the author's experimental and numerical computational works carried out at the School of Mechanical Engineering at the University of New South Wales of Australia with a particular focus on the development and production of highly efficient wind driven rotating ventilators for domestic and industrial use.

1.1. Significance of Ventilation

In the past, the building codes and policies of most of the developed nations were implemented to ensure the safety and adequate sanitary conditions for occupants and protect the public from hazards such as fire, electrical problems, sewage disposal, etc. It is more recently that emphasis has been placed on energy efficiency and sustainable buildings. Issues have varied over time, but recently the catalyst for such policy changes has come about as a result of rising greenhouse gas emissions from energy consumption, as well as land use practices such as deforestation. Thus governments around the world are grappling with ideas regarding building policies and the potential benefits to not only the people that use them but the environment in which the buildings exist.

Generally speaking, a movement or circulation of the air within an enclosed space in any ventilation system is essential to ensure that the temperature and humidity be maintained within a range that allows adequate evaporation of perspiration from the skin. There is evidence to show, that the deleterious effects result largely from interference with the heat-regulating mechanism of the body. Lack of air currents and the increase in relative humidity and temperature, especially noticeable in crowded or poorly ventilated places, prevent normal evaporation of perspiration and loss of heat from the surface of the skin. Under requirements to maintain a safe working environment, many dwellings and factories now need adequate fresh air exchange to remove gaseous, process emissions and/or heat buildup. The high priority placed on indoor air quality from health considerations has prompted New York in the USA to pass legislation effective from December, 2008 to require landlords to notify tenants and building occupants of indoor air test results.

Good ventilation is a key element in maintaining a healthy environment for people within the confines of a building. Failure to provide adequate ventilation in a building can result in problems with moisture, unpleasant smell, lack of oxygen, and unacceptable content of poisons gases such as CO which cause medical conditions for tenants. Contaminants such as formaldehyde or radon can also accumulate in poorly ventilated homes, causing health problems. Due to fact that every person's resistance to pollutant varies, it is hard to obtain an accurate quantitative measurement of the impact of ventilation on human beings (Hanssen 1997). However, some method was proposed based on the consideration of the factors such as direct medical costs and lost earnings due to major illness as well as increased employee sickness days and lost productivities while on the job (European Collabrative Action report 2000). For example, Brooks and Davis (1992) calculated that productivity loss in the United States of America attributed directly to indoor air quality (IAQ) is around 14 minutes per day;Pillgram Larsen (1991) estimated loss in Norway is of 1 to 1.5 billion of euro per year which is equivalent of 250 to 350 euro per inhabitant. For Finland, a more detailed estimation was given by Seppanen and Palonen (1998) where the potential economic impact of indoor air pollution was estimated to be tens of billions of dollars per calendar year (see Table 1.).

Table 1. Cost due to poor indoor air quality

Building Type	Maintenance Costs (Billion Finnish Marks)	Energy Costs (Billion Finnish Marks)	Costs of poor indoor climate (Billion Finnish Marks)
Houses	12.7	7.2	7.0 (Mould allergies)
High-rise residential building/office buildings/schools/hospitals and other buildings	37.3	7.0	10.6 (Radon cancers, decreases in efficiency, poor learning, hospital infections)

Ventilation is also an effective means of ameliorating the internal environment (Billington 1982). This is usually achieved in three prime motive forces: mechanical means(fans, pumps, HAVCS), heat (stack effect) and wind pressure. The use of mechanical ventilation system consumes a significant amount of energy in a building. For example, electricity bills for residential buildings in the United States of America is approximately US$140.8 billion in 2006 (U.S. department of energy 2006), and the energy consumed by heating, ventilating and air conditioning system (HVAC) accounts for 40 to 60 percent of the total energy. For example, the energy cost has risen 30% in the state of New South Wales since 2009 (IPART website 2009). Due to the rising fossil and energy costs, people's awareness have increased and attention is focusing towards the use of an alternative solution. Under such concern, natural ventilation becomes a viable and sustainable method in providing suitable environment and is increasingly considered in many building designs.

2. BRIEF OVERVIEW ON DIFFERENT MODES OF VENTILATION

In general, natural ventilation can be induced by cross ventilation or stack ventilation and they can be either passive or active in nature. Cross ventilation technique use the natural wind force to direct the air movement through a building. When wind impact on a building it produces positive pressure on the windward side and relative negative pressure on the leeward side and these difference in pressures drive the airflow through the building. In order to achieve good cross ventilation effect, openings such as windows, doors and other openings are required at the different facings of the building. The stack ventilation technique utilizes the temperature between inside and outside of a building. When the room temperature is greater than outside, the warm indoor air will rise and exit and cooler, denser air from below can enter.

The natural ventilation can be further enhanced by wind driven ventilation technique. These techniques can be either passive or active in nature however the primary driving force for providing ventilation are primarily caused by wind.

2.1. Passive Wind Driven Ventilation Techniques

2.1.1. Window Openings/Vents

This is probably the simplest form of its kind. Through strategically placement of openings, the potential natural ventilation in a building can be optimized to cut down the need for artificial cooling (Mochida et. al. 2005, Karave et. al. 2005). In conjunction with the window openings, static vents can also be placed at the planned location where window openings are not required. There are various of types of vents that are commercially available as shown in Fig. 1. Due to the relative cost effective and sustainable benefits, a large number of experimental and numerical investigations have been conducted to study the benefit of this technique (Greeno 1997, Heiselberg et al., 2001, Olsen & Chen 2003, Faye et. al. 2005, Asfour & Gadi 2007).

The low-tech nature and manual operability of window openings/static vents technique are acceptable in domestic dwellings. However, a major limitation of naturally ventilated technique in general is the unpredictability of the driving force. Wind pattern and availability for different region varies according to seasons and more over for large buildings some physical limitations exists. Hence its application on its own cannot be considered to be a primary ventilation method.

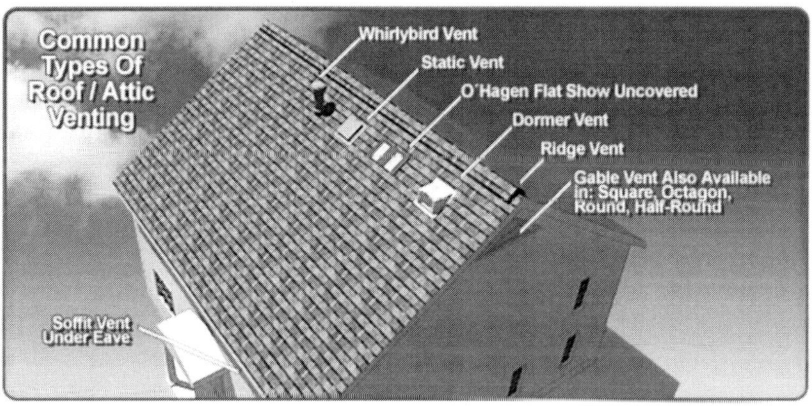

Figure 1. Common types of roof/attic venting

2.1.2. Atria and Courtyards

Courtyard is an effective way of providing natural ventilation and can be found in many architectural designs across the globe for thousands of years. A courtyard provides a relatively enclosed space to channel and direct airflow into some openings and results in convective natural ventilation within and around the building. Some studies have been conducted to study the effect of atria and courtyards on energy performance (Rajapaksha et. al. 2003, Aldawoud & Clark 2008). These studies generally confirmed that the wind driven ventilation can be enhanced by using atria and courtyards and are more effective for low rise building. However, the passive nature of this technique indicates that its effectiveness depend largely on the availability of its driving force.

2.1.3. Wind Towers

Wind towers have been used in conjunction with courtyard and atria for centuries in many middle-eastern architectures. They work on the principle of both wind driven and stack effect ventilation. Wind enters via the windward side that has a positive pressure coefficient and was redirected through the tower into the building. These winds disperse through the building and exit through relative negative pressure areas. The exit paths can be multiple, either through the tower or openings that was carefully designed. Multiple wind towers have also been used to maximize the effect as can be seen from Fig. 2. Wind towers can also benefit for modern buildings and have been the subject of several studies (Bansal et. al. 1994, Bahadori 1994, Kolokotroni et. al. 2002, Badran 2003) through the following considerations: (1) Tower with evaporated cooling columns can increase the cooling potential of incoming air. (2) Natural ventilation principle can dictate the extract height and cross section of the towers to accommodate for the targeted number of occupants. (3) Solar collectors can be used to enhance the stack effect ventilations in times or areas of little wind.

A modern modification of wind towers is a device known as the wind catcher. These devices capture the wind at roof level and direct it down to the building. The working principle and their application of these devices are shown in Fig. 3.

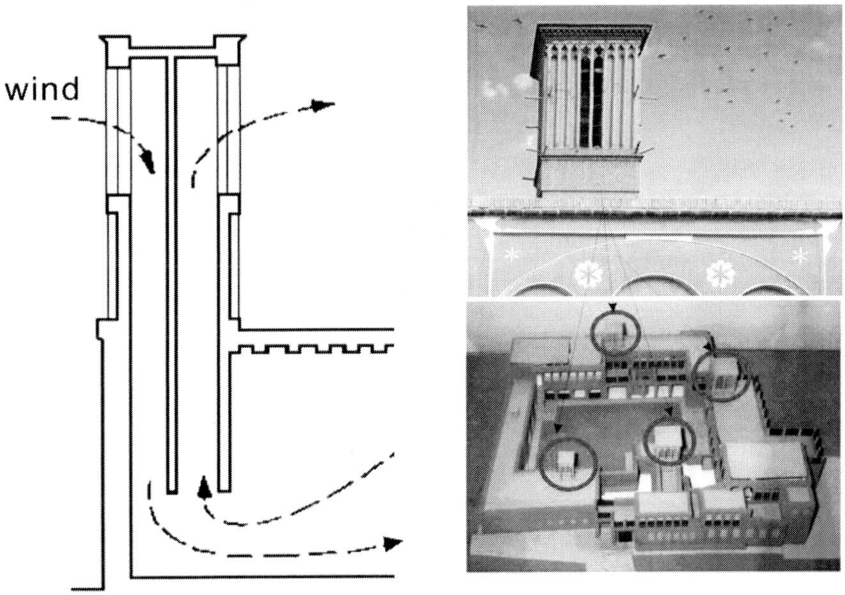

Figure 2. Wind tower design in Middle Eastern building

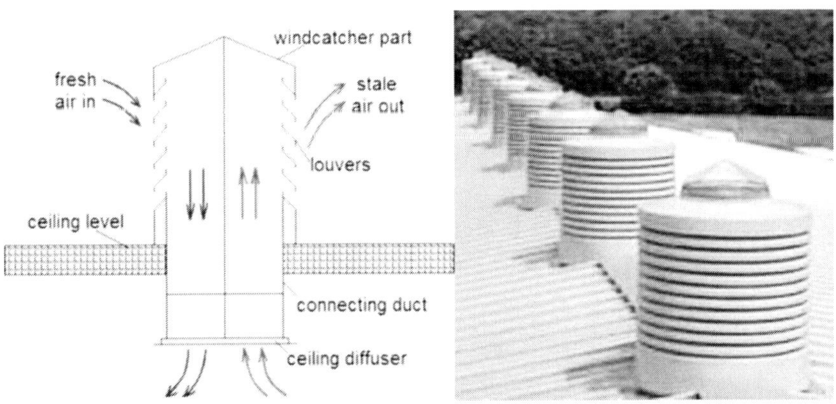

Figure 3. Operation principle and application of wind catcher system

Since the introduction of this technique in 1990 there have been a number of studies associated with its performance both experimentally and numerically (Kolokotroni 2002, Li & Mak 2005, Elmualim 2006). These studies points put that the performance of a wind catcher system is influenced by a number of factors: the outdoor wind; temperature difference between indoor and outdoor environment and location of other openings and windows.

The studies surrounded the wind tower technique not only indicates that the principle works, more importantly they highlighted the needs of redundancy via another source when using wind driven ventilation technique such as active system.

One major concern of the passive natural ventilation technique is that the driving forces are variable and the minimum required ventilation rate is hard to be ensured in still air conditions (no wind or stack effects). Hence several alternative engineering solutions are formulated to maximize and control the effect of natural ventilation. For example, the roof mounted rotating turbine ventilators can be used or in conjunction with the natural ventilation techniques. The rooftop ventilators are commonly found in most countries of the world where they are mounted on the roof and rotate with the slightest breeze and generate a powerful updraft to effectively pump the air out of the room.

2.2. Active Wind Driven Ventilation Techniques

2.2.1. Turbine Ventilators

Turbine ventilators are available commercially with various blade designs, sizes and materials. They usually rotate in its vertical axis to create updraft inside the turbine which extracts air and in the absence of wind, they can still facilitate ventilator using the stack effects. These types of installations are usually on the top of the roof where higher wind speeds are available for most buildings and these are shown in Fig. 4.

Figure 4. Various blade design of turbine ventilators

The concept of a turbine ventilator was originally patented in 1929 by Meadows and it's commercial production was started by Edmonds of Australia since 1934. Although these types of devices have been on the market for many decades, there were only limited quality aerodynamic researches available in open literature.

The first series of investigation into the aerodynamics of a turbine ventilator was reported by Lai (2003). His flow visualization study on three different seized turbine ventilators of 6, 14 and 20 inches respectively showed the airflow pattern around a turbine ventilator. Another experimental study byFlynn and Ahmed (2005) also confirmed the flow pattern. These studies suggested that the larger diameter ventilators induce greater mass flow extraction rates.

Figure 5. Installation of multiple stacked ventilators at UNSW

The performance studies of a turbine ventilator are usually associated with the air extraction rates (Revel 1998, Khan et. al. 2008). These studies suggested that the performances of such devices are significantly influenced by the blade design as well as the construction material and the sizes significantly affect producing higher exhaust mass flow rate. On the other hand, the aerodynamics forces acting on a turbine ventilator have also been studied by Rashid and Ahmed (2003). It was concluded that the operation of a ventilator without the effect of inclined rooftop is more efficient at lower wind speeds. At higher wind speeds larger flow separation was induced on the blades and this reinforced the need of greater attention to optimize blade designs so that they are capable of operating over a wide range of wind speeds.

2.2.2. Hybrid Ventilators

As the performance of the traditional turbine ventilators largely depend on the local wind conditions and the ventilation rates are cannot be guaranteed, hence various of forms of hybrid system are purposed to boost its performance to comply with the increasing building regulation. In this section only three examples of ventilation systems that rely purely on wind power or renewable power source are discussed.

The first such ventilator system considered has been proposed by Lai (2006). This ventilator combines the traditional turbine ventilator with a small DC fan in the base duct which is powered by solar cell as presented in Fig. 6. This system was found to be effective in enhancing the ventilation rate at low wind speeds.

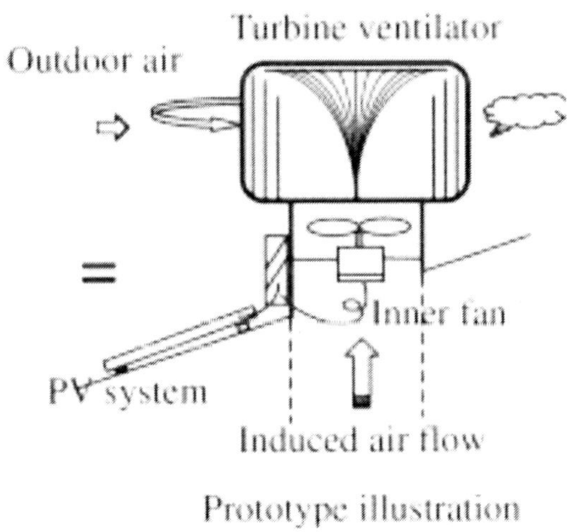

Figure 6. Prototype turbine Ventilator (Lai, 2006)

The second ventilator system considered uses a different approach. It can be easily demonstrated that any obstruction in the throat of a wind ventilator can produce significant noise and reduce vent performance, and this performance reduction can be as high as 40% (West 2001, Khan et. al. 2008) indicates than under normal wind load. The UNSW research team under an Australian Research Council Grant in active collaboration with CSR Edmonds, Pty Ltd, therefore, proposed a model (Ahmed, 2010), the EcoPower, which has no motor and fan blade in the throat of the vent. The model utilises an electronic commutating (EC) motor installed in the head of the ventilator to enable motorised boost during periods of low wind speed or special ventilation needs. The motor can also be programmed and activated by digital measures, such as temperature, humidity gas concentration level etc or by manually switch. The research behind this device also indicates a significant improvement of ventilation rate at low wind speeds. The EcoPower is now commercially produced and marketed worldwide by CSR Edmonds, Pty Ltd.

The third approach mentioned here is the hybrid system proposed by Arup. This device incorporates a vertical axis wind turbine that produces renewable energy to drive the fan which is located directly underneath the turbine to produce the desired ventilation. A photograph of this device is presented inFig. 8. This first full scale model was constructed and installed at the Harrare International School in Zimbabwe. It was found that the room temperature in the classroom is significantly cooler (up to 80º C) during the day than without the system installed. However there is very little academic literature on the flow rates and performance associated with this device.

3. TOWARDS ENHANCED WIND DRIVEN VENTILATION TECHNIQUE

In this section some recent studies on means to enhance the performance of wind driven ventilator by the authors at University of New South Wales have been summarised.

Figure 7. EcoPower Ventilator (CSR Edmonds, Pty Ltd)

Figure 8. Vertical axis Wind Extractor (Arup, Zimbabwe)

3.1. Physical Experimentation

The physical experimentation were conducted in a 76mm diameter open return, 0.2% turbulence intensity open test section wind tunnel of the Aerodynamics Laboratory of the University of New South Wales. The experimental setup is shown in Fig. 9.

A five-hole pressure probe was used in this investigation. Recent advances in using multi hole pressure probe techniques (Pisasale & Ahmed, 2002-2004), makes it possible to obtain the velocity vector and static pressure field in a highly three dimensional flow field. A step by step procedure of how to obtain the velocity vector and pressure terms can be obtained is detailed in reference 21. A Furness FCO510 micromanometer and a box consisting of seven channels of Honeywell transducers were used to measure the pressures on the five-hole pressure probe. The manometer and the pressure transducer provided measurement in the range that varied from -2000 Pascal (Pa) to 2000 Pa with an accuracy of ±0.1Pa.

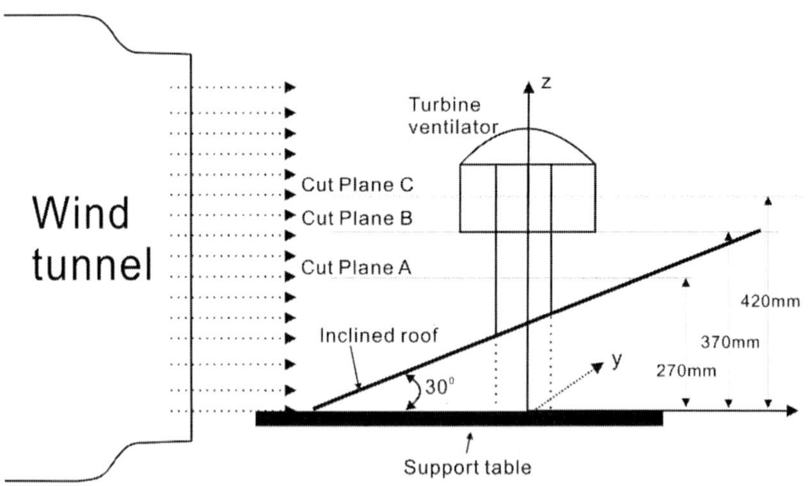

Figure 9. Schematic of experimental setup

Apart from the forces obtained directly from the force balance, pressure and skin friction distribution were used to determine a flow features of the system. Since these parameters contribute largely to the aerodynamic forces acting on a body. In the present investigation, the effects of the inclination angles on the performance of a rotating ventilator were investigated in a wind tunnel. The Reynolds number, Re_x (= xU_∞/υ), used was based on, x, the distance measured from the leading edge of the plate, $U_{\infty,}$the free stream velocity and υ, the kinematic viscosity of the fluid. Re_x values ranged between $7 \times 10^4$and 1×10^6.

3.1.1. Effect Of Inclination Angle On Forces on a Ventilator

The force components in the x, y and z directions as well as the total force acting on the rotating ventilator at different inclination angles were examined against free stream velocities and presented inFigures 10a to 10d.

From these Figures, the force acting on the turbine ventilator is found to increase with free stream velocity. The increasing trend is more pronounced in the x direction signifying that the inclination angle has greater effect on F_x which largely contributes to the drag than on the side and normal components of the force that contribute to both lift and drag (Rashid & Ahmed, 2003).

Some additional features may be elucidated if the total and its component forces (F_t, F_x, F_y, F_z) are examined in non dimensional forms or force coefficients (C_t, C_x, C_y, C_z) against Reynolds number of the flow (Re_d) based on the rotor diameter. The total force and its components were normalised by $0.5\rho U_\infty{}^2 A$, where ρ is the density of air, U_∞ is the free stream velocity and A is the total wetted area of the ventilator. The wetted area was calculated based on the total area exposed to the flow (see Fig. 9.). The results are shown in Figures 11a to 11d.

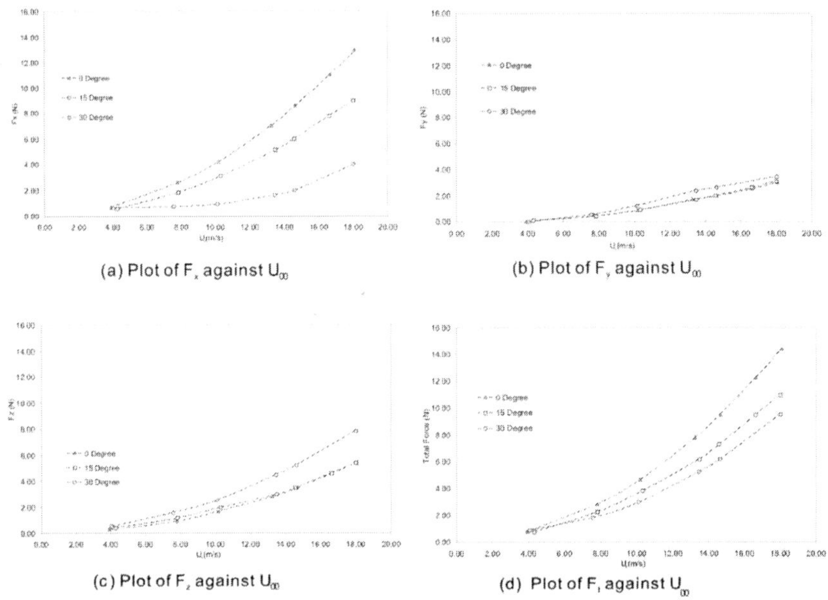

(a) Plot of F_x against U_∞

(b) Plot of F_y against U_∞

(c) Plot of F_z against U_∞

(d) Plot of F_t against U_∞

Figure 10. Force acting on the turbine ventilator at various inclination angles.

It is evident that the C_x value decreased with the increase in inclination angle while the C_z value showed the opposite trend. The inclination angle, however, showed little or minimum effect on the $C_y$values. Consequently, the total force coefficient, C_t, acting on the ventilator reduced with increases in inclination angle and converged towards a constant value beyond Reynolds number of $2x10^5$. This is significant as it implies that the force acting on the ventilator is independent of the Reynolds number beyond a certain rotational speed and may be highly relevant during the operation of a turbine ventilator at lower wind speed. A hybrid system that incorporates other power source such as solar power may provide additional means to efficiently overcome the inertia and friction forces acting on the body to initiate adequate rotation at low wind speed.

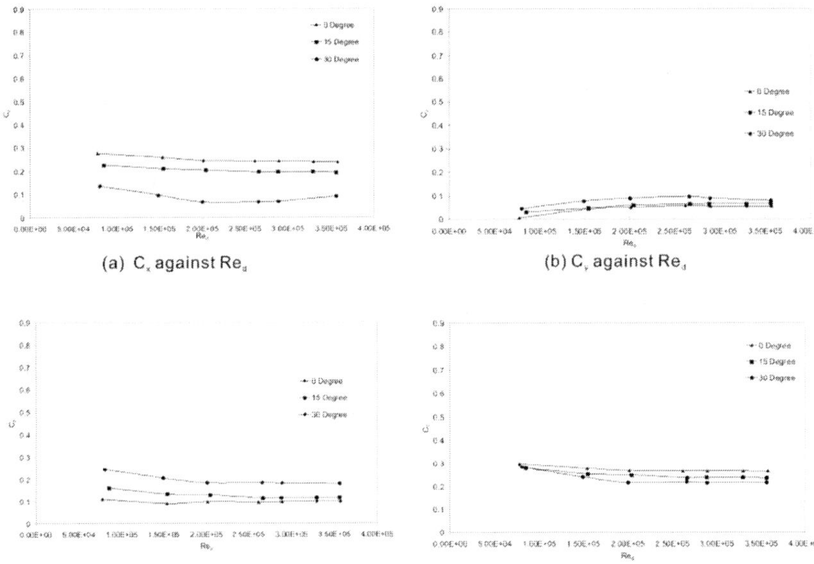

(a) C_x against Re_d (b) C_y against Re_d

(c) C_z against Re_d (d) C_t against Re_d

Figure 11. Coefficient of force at various inclination angles.

3.1.2. Effect Of Inclination Angle on Rotational Speed Of A Ventilator

The overall effect of the roof inclinational angle on the operation of a turbine ventilator was further examined against its rotational speed. The results are shown in Fig. 12. The results indicate that the rotation speed of the turbine ventilator showed a linear relationship with the free stream velocity (Flynn & Ahmed, 2005; Rashid & Ahmed,2003; and Khan et al., 2008). Additionally, it also indicates that the high inclination angle lowers the rotational speed of the ventilator that is more noticeable at higher free stream velocity suggesting that the inclination angle has minimal impact on the total mass flow extracted at lower wind speeds. This is also a significant observation since the operation of these ventilators are designed for low wind speeds and as such will not suffer greatly from adverse effect of the high inclination angle.

The Australian regulation stipulates that for the safety of operation, the turbine ventilator must be able to withstand wind loads from speeds of 200km/hr without suffering deformation. As such loads are directly proportional to ventilator rotation, at high wind speed safety in operation become more of a priority than mass extraction rate. Thus, the lowering of rotation due to increasing inclination angle of roof may thus be usefully exploited to reduce wind loads and hence extend the safety margin.

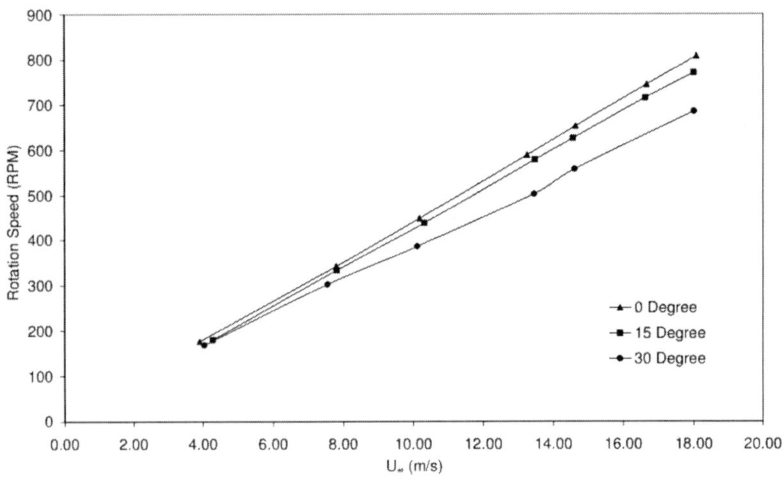

Figure 12. Rotation speed of the turbine ventilator at various inclination angles.

3.1.3 Normalised Static Pressure Distribution on the Roof

The results of static pressure distribution in non-dimensional form, Cp, under various inclination angles on the inclined roof without the turbine ventilator are presented in Figures 13a to 13c, where Cp = (p_s-

$p_\infty)/(0.5\rho_\infty U_\infty{}^2)$, p_s is the static pressure measured at each static ports on the surface and p_∞ is the static pressure of free stream flow. The results obtained when the turbine ventilator present are shown in Fig. 13d to 13f. An advantage of such representation is that the Cp values can range from +1 to any negative value. A positive Cp values generally suggests that the flow is slowing down until reaching the limiting value of +1 or the stagnation or zero velocity condition, while any negative Cp value means that the flow is experiencing acceleration, with higher negative Cp signifying higher acceleration.

The inclined roof at its leading edge imposed pressure gradient to the oncoming flow and forced it to slow down producing a Cp distribution that had values moving towards +1 with approximately a symmetrical distribution in the spanwise direction. The Cp values at the leading edge for 0, 15 and 30 degrees of inclination were found to be 0.05, 0.19 and 0.42 respectively that decreased towards zero value in the streamwise direction.

On the inclined plane, the addition of the ventilator imposed further pressure gradient that slowed the incoming flow even further. From Figures 13d to 13f, the Cp values at the leading edge increased to more positive Cp values due to the presence of the ventilator while maintaining symmetrical distribution for the upstream flow as expected. Similar trends were also observed for flows at 5m/s suggesting that the Reynolds number appeared to have negligible effect on the overall pressure distribution.

3.1.4. Skin Friction Distribution on The Roof

Recent advances (Pisasale & Ahmed, 2002-2004, Lien & Ahmed, 2003) in using multi hole pressure probe have greatly aided in the quest of measuring the velocity and surface skin friction distribution in a complex flow field. Following the procedure outlined in Lien & Ahmed (2003), a five hole pressure probe was used to measure the skin friction distribution at each measuring station.

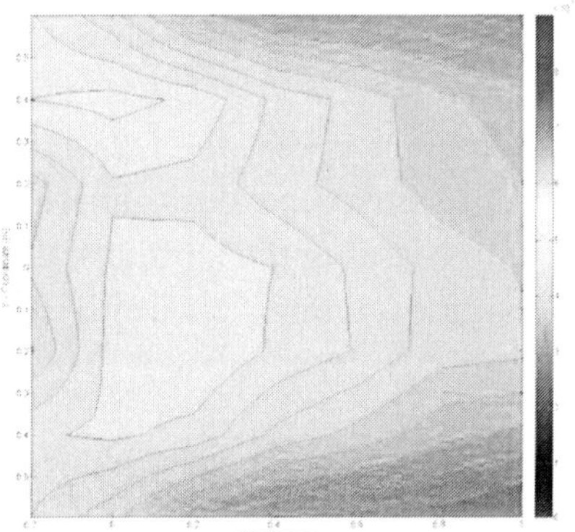

(a) 0 Degree (inclined roof only)

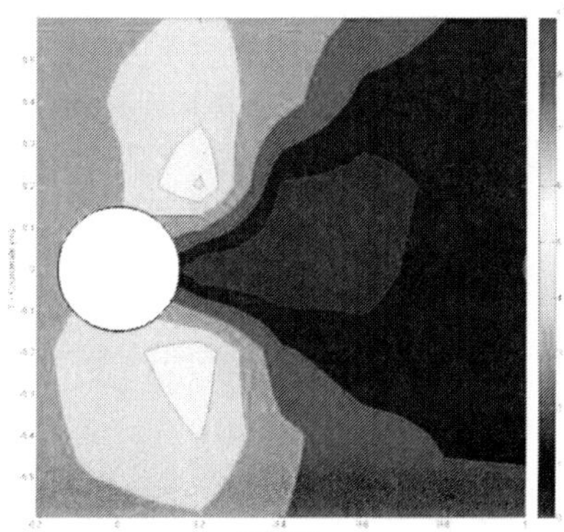

(b) 0 degree (with rotating ventilator)

(c) 15 Degree (inclined roof only)

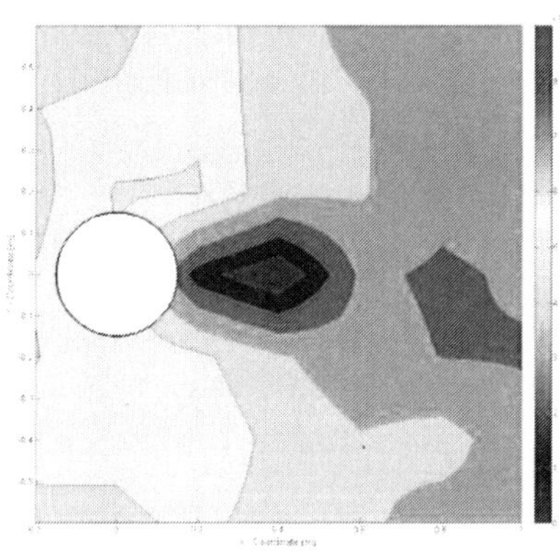

(d) 15 degree (with rotating ventilator)

(e) 30 Degree (inclined roof only)

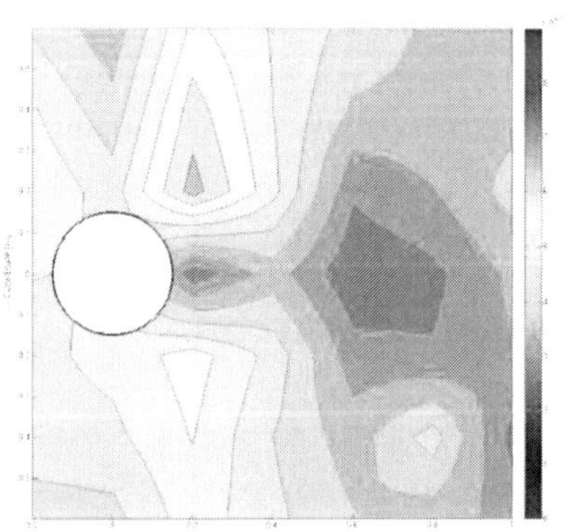

(f) 30 degree (with rotating ventilator)

Figure 13. Static Pressure distribution on the roof, where $U_\infty = 10$ m/s

The skin friction distributions on the roof at different inclination angles 'without the ventilator' are shown in Figures 14a to 14c and 'with the ventilator' present are shown in Fig. 3.6d to 3.6f.

From tests conducted at the two wind tunnel speeds of 5m/s and 10m/s and at the three different inclination angles of 0, 15, and 30 degrees, the skin friction distributions were found to become progressively lower downstream of the ventilator. This reduction in skin friction values appears to be a consequence of reduced free stream velocity due to the momentum deficit occurring behind a bluff body. Since skin friction varies with square of the free stream velocity (Baskaran & Bradshaw, 1993), it is not surprising that the trend is more apparent at higher wind speeds.

On the other hand, with the ventilator placed on the roof, the skin friction distribution on the roof increased significantly with increases in the roof inclination angle. Since the skin friction signifies losses in the flow, it is expected that the increase of this quantity appears to emanates from the overall reduction of the forces acting on the ventilator that produces lower rotational speed.

3.2.Cfd Modelling

After gaining insights of flow physics around the ventilator and on the inclined roof, Computational Fluid Dynamic software is also utilised to extend the performance study of rooftop ventilator under the effect of inclined roof.

The turbine ventilator used in this investigation is a Hurricane H100 produced by CSR Edmonds Australia Ltd. It consists of a rotating portion (Rotor) with 8 curved blades and a stationary portion in the form of a cylindrical base. The dimensions of the various components of this ventilator are shown in Fig. 15.

This numerical simulation of flow characteristics of the turbine ventilator used in wind tunnel tests were then performed using

FLUENT. The equations for the conservation of mass, momentum and turbulence scales are solved in FLUENT using the control volume method in a three-dimensional body-coordinate system.

The geometry is created in Computer Aided Three-dimensional Interactive Application (CATIA) and exported to GAMBIT (the pre-processor) for meshing. For modelling of the rotating motion of the turbine ventilator, the multiple reference frame (MRF) meshing in FLUENT was adopted as they have been successfully applied in flows with rotating objects (Luo et al., 1994). This meshing allows multiple moving reference frames to be solved in a single domain. MRF is normally used for steady state condition and the individual cell zone can move at different rotational speeds and provide solution using the moving reference frame equations. Grid sensitivity test was performed by examining the effect of different number of mesh grids and the total number of grids was determined when grid independence was established. The final total number of grids used in the present simulation was determined to be approximately 1,200,000. The maximum and minimum grid volumes were approximately 1.8×10^{-4} m^3 and 2.4×10^{-9} m^3 respectively. Mesh quality was checked in GAMBIT to ensure the skewness of the three dimensional grids was less than 0.8.

3.2.1. Performance Investigation of Rooftop Turbine Ventilator

The major function of a rooftop turbine ventilator is to create suction and move air out from an enclosed building. The performance of a wind driven ventilator is generally gauged against its exhaust mass flow rate. However, before exhaust flow rate was determined, an attempt was made to explore the nature of the flow associated with the rotating ventilator particularly those inside the rotor.

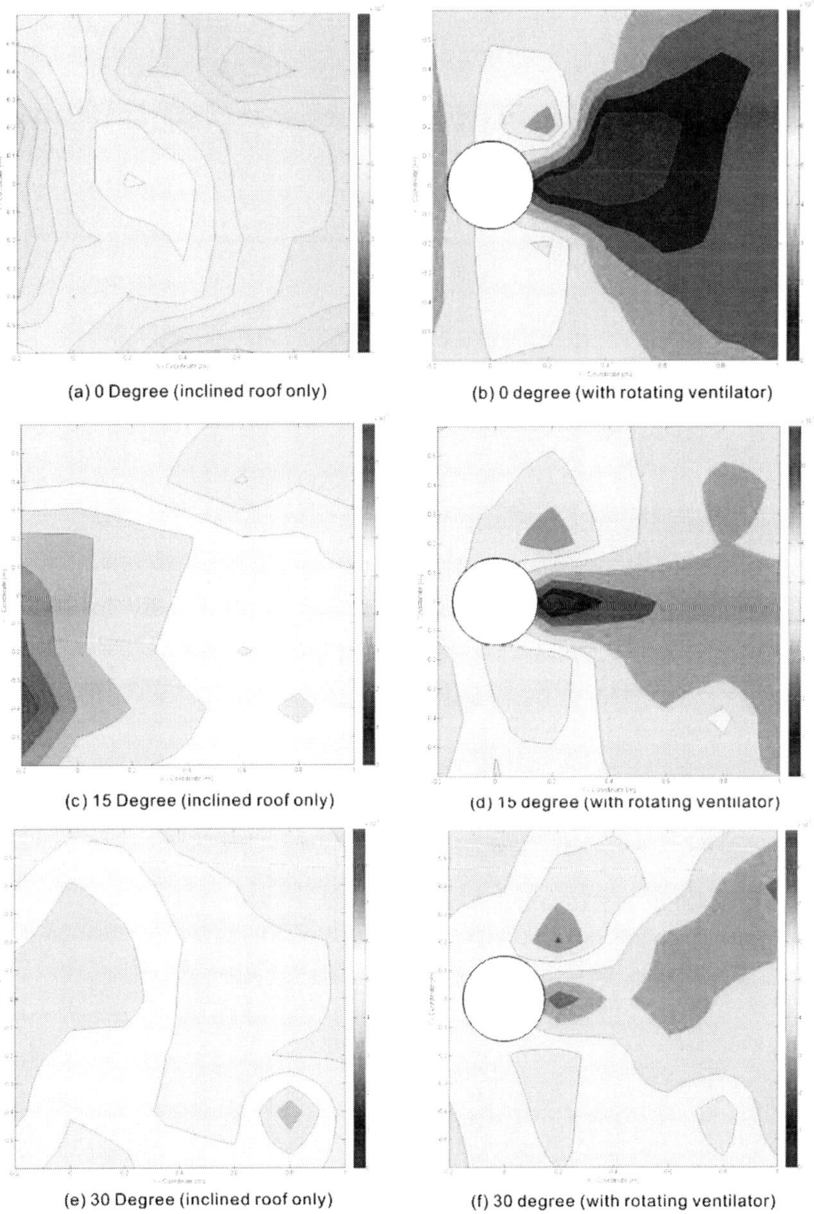

(a) 0 Degree (inclined roof only)
(b) 0 degree (with rotating ventilator)
(c) 15 Degree (inclined roof only)
(d) 15 degree (with rotating ventilator)
(e) 30 Degree (inclined roof only)
(f) 30 degree (with rotating ventilator)

Figure 14. Skin friction distribution on the roof, where $U_\infty = 10$ m/s

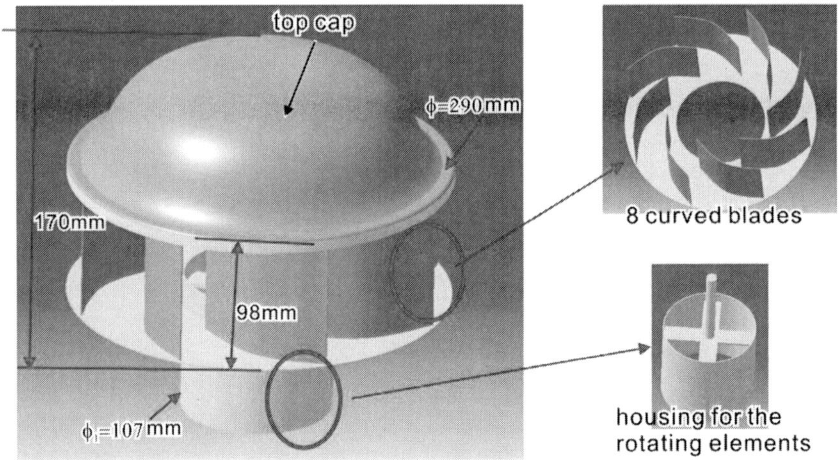

Figure 15. Modelling of turbine ventilator

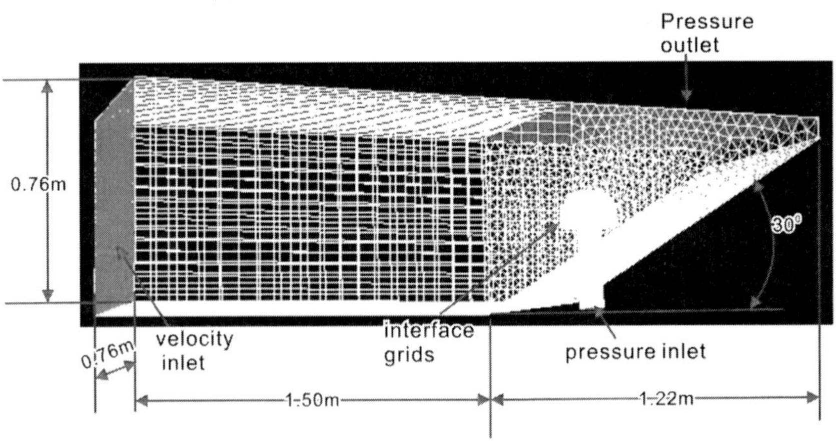

Figure 16. Modelling of domain and dimension

CFD results can provide clear image of the external and internal flow on the entire ventilator in terms of the three dimensional path lines. This provided a qualitative description of the general flow pattern. This is shown in Fig. 17. Two distinct streams of the flow can be observed. The left hand-side flow follows the direction of the turbine rotation while the flow direction of the right-hand side flow is contrary to the turbine rotation. The right hand side flow becomes the defected flow field that acts against the rotation of the turbine. This results in the formation of secondary circulation between the separated flow and the blades on the right-hand side flow.

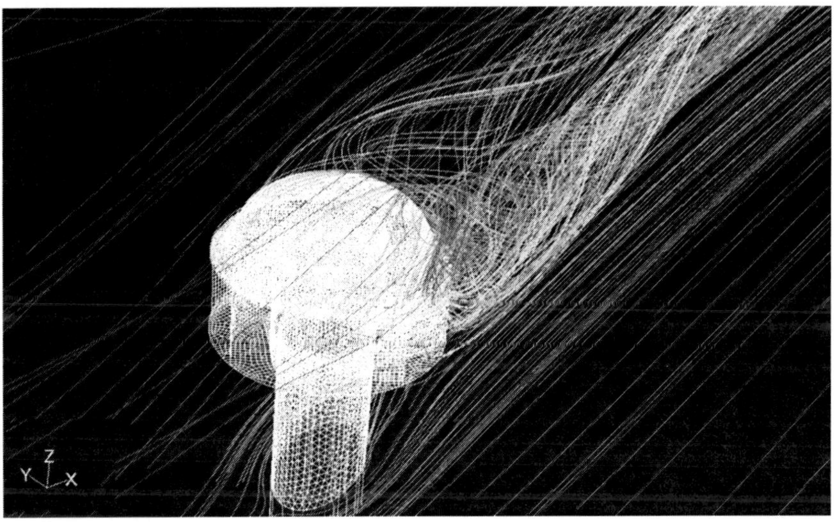

Figure 17. Three dimensional path line of the flow associated with the rotating ventilator at 10m/s.

The CFD software can further observe the features of the behaviour of the flow as they are extracted and exhausted by the ventilator as shown in Fig. 18. The path lines of the flow at different wind speeds within the rotor and downstream in the wake are shown. The rotor cap is removed from this Fig. for greater clarity. It is clear that at various wind speeds tested, the rotation of the turbine creates updraft.

The flow interactions within the rotor can be shown using vector plots in three dimensions and the results are shown in Fig. 19a to 19c. The path line in the duct (base) and the rotor also indicates that the flow swirls upwards and mixes with the secondary circulation of the wake region downstream of the ventilator.

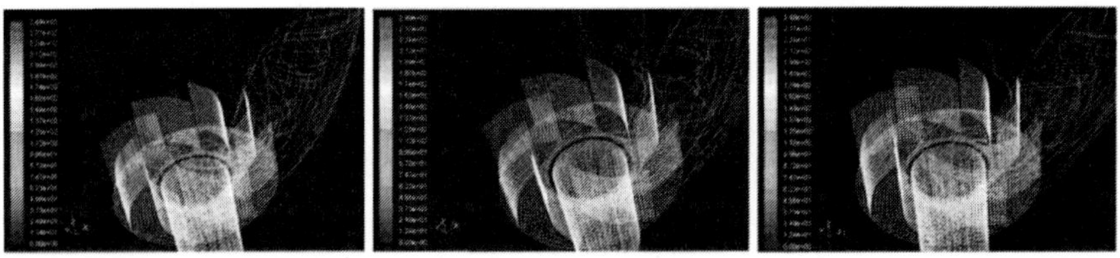

(a) Free Stream Velocity = 5m/s (b) Free Stream Velocity = 10m/s (c) Free Stream Velocity = 15m/s

Figure 18. Three dimensional path lines of flow within and downstream of the rotor.

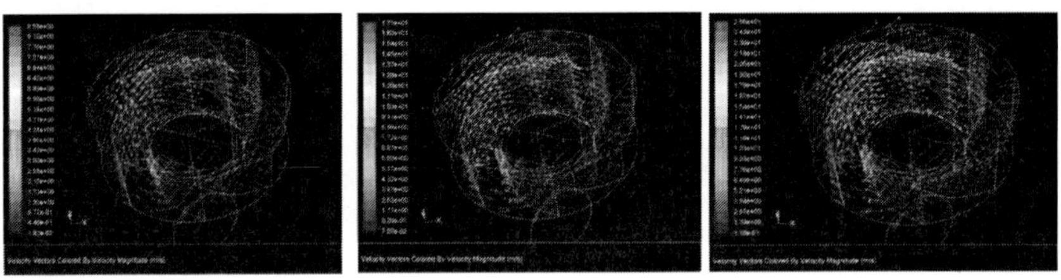

(a) Free Stream Velocity = 5m/s (b) Free Stream Velocity = 10m/s (c) Free Stream Velocity = 15m/s

Figure 19. Velocity vector inside the ventilator

It is evident that as the ventilator rotates in the clockwise direction, the air flow extracted through the base swirls up and mixes with the oncoming wind entering the ventilator from the left. The resulting mixed air exhausts from the right half of the ventilator. These Figures also show that with increasing wind speed there is a corresponding increase in swirl flow within the rotor suggesting subsequent increase in exhaust mass flow rate.

To provide further insight, the flow features within and around the rotor at the two cut planes of B and C at three wind speeds are presented in Figures 20 and 21 respectively. The results are given as two dimensional vector plots.

Figures 20 and 21 indicate that with increasing wind speeds, the flows at the left side of the ventilator are higher than at the right side. The extracted air swirls in clockwise direction. Figures 21a to 21c further indicate that the wake behind the rotor increases in size with increases in wind speeds that produces the increases in suction and hence the increases of swirl component within the ventilator. These observations are in good agreement with the flow visualisation experiments conducted by Flynn &Ahmed (2005) and Lai (2003).

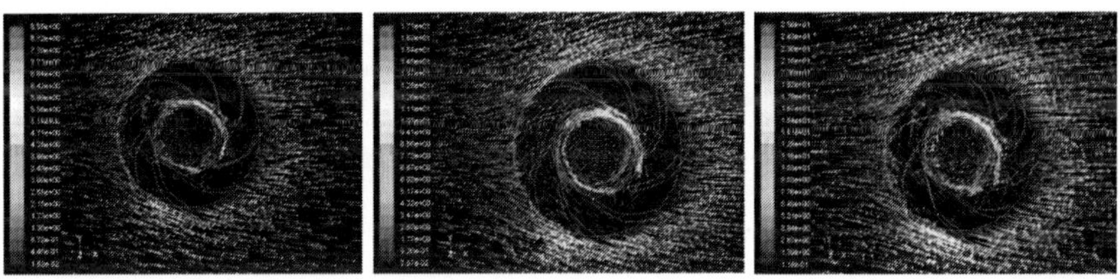

(a) Free Stream Velocity = 5m/s (b) Free Stream Velocity = 10m/s (c) Free Stream Velocity = 15m/s

Figure 20. Velocity vector field at cut plane B at different wind speeds

(a) Free Stream Velocity = 5m/s (b) Free Stream Velocity = 10m/s (c) Free Stream Velocity = 15m/s

Figure 21. Velocity vector field at cut plane C at different wind speeds

To explore the capability of the CFD modeling approach as an initial design tool, the exhaust flow rate was determined against a range of wind speeds between 3m/s to 15m/s. Numerical computation were performed by assigning boundary condition of 'pressure inlet' at the bottom of the pipe set as ambient pressure.

Once the exhaust mass flow rates for various cases were obtained, the effect of blade heights on performance of a turbine ventilator was explored. Three blade heights H, 1.5H and 2H were considered, with H = 98mm. The computed exhaust mass flow rates for varying wind speeds are shown in Fig. 22.

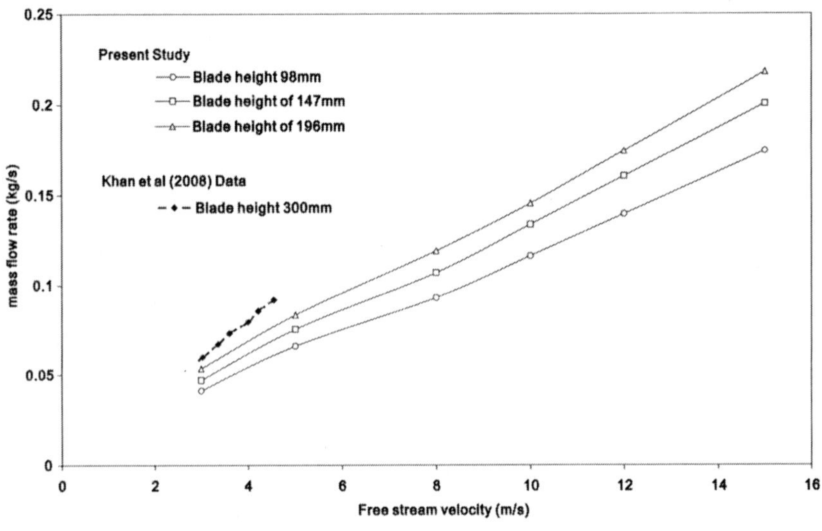

Figure 22. Effect of free stream wind speed on exhaust mass flow rate

It is clear that the exhaust mass flow rate has a near linear relationship of increasing in value with increases in the incoming wind speed. The same trend has been observed by Khan et al., (2008) where four different geometry ventilators were tested in a wind tunnel.

Additionally, the wind tunnel results of Flynn & Ahmed (2005), Khan et al., (2008) indicate that the rotational speed of the ventilator also increases linearly with increases in wind speed suggesting a linear increase in mass flow rate with increasing ventilator size (diameter or blade height). With 50% and 100% in the ventilator blade height, the improvement of exhaust mass flow rate was found to be 15% and 25% respectively.

Due to the absence of adequate performance data on turbine ventilator in open literature, the data from the experimental results of Khan et al., (2008) on Hurricane turbine ventilator of 300mm diameter and 300mm blade height were extrapolated to provide some comparison against the CFD results. The results are presented in Fig. 23.

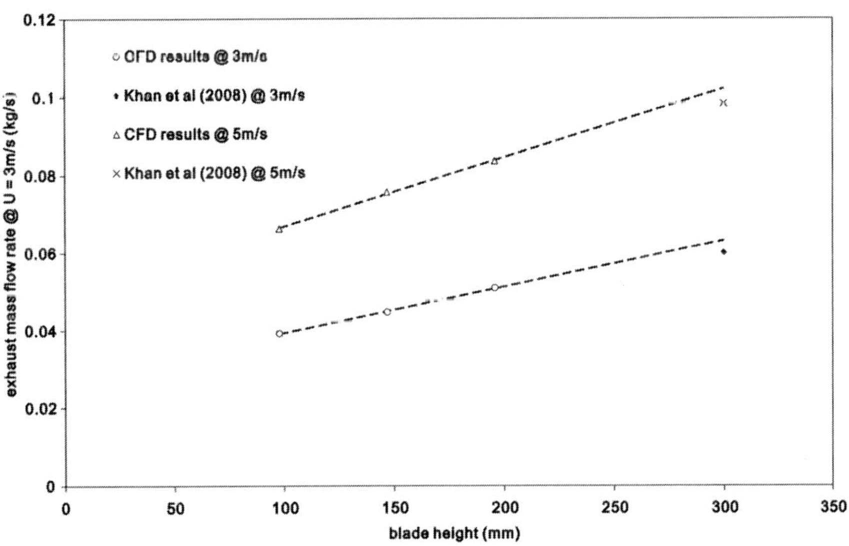

Figure 23. Effect of blade height on exhaust mass flow rate (U_∞= 3m/s and 5m/s)

Again from Fig. 23, a linear relationship between exhaust mass flow rate with blade height was observed. A slight overprediction when compared against Khan et al.,(2008)'s data was observe. Given that the comparison was with extrapolated value and also for the fact that CFD results were from an inclined roof, the small discrepancies are not surprising. However, based on the observations that the similar trends were observed for the exhaust mass flow rate as well as the external and internal flows associated with a rotating ventilator on a 30^0 inclined roof, it appears that the roof inclined at such angle did not have a significant effect on the overall trend in ventilation performance of a rotating ventilator.

4. CONCLUSIONS

Wind driven ventilation systems that utilize wind as a natural energy to provide improved air quality within buildings have been considered in this chapter. Since the wind driven turbine ventilator has proved to be one of the most efficient such device, it has been the primary consideration in this chapter. However, the performance of wind driven ventilation depends largely on prevailing winds. This has necessitated efforts of the authors to concentrate on exploring the basic flow physics around and within such wind driven ventilator that are detailed in this chapter. The results indicate that the efficiency and reliability of a ventilator operation may require boosting by a hybrid system by incorporating other power sources such as solar power to overcome the inertia and friction forces acting on the body to initiate ventilator rotation at lower wind speed.

Numerical analysis using computational fluid dynamics (CFD) was also employed to investigate both internal and external flows around and within a rotating ventilator. A promising conclusion that can be drawn from these studies is that CFD analysis could be used alongside physical experimentation, as a cost effective aid to enhance operational performance and safety features in the future design and development energy efficient natural wind or wind/hybrid powered ventilation systems to improve air quality and comfort within buildings.

5. ACKNOWLEDGEMENTS

The authors would like to acknowledge the Australian Research Council and CSR Edmonds Pty Ltd, Australia for providing the funding; Managing Director, Mr. Allan Ramsay and National Technical Director, Mr. Derek Munn, of CSR Edmonds and Technical Officer, Mr. Terry G Flynn, of Aerodynamics Laboratory of University of New South Wales for their assistance during the conduct of this study.

REFERENCES

1 Ahmed, N.A. (2010). Wind-Solar Driven Natural Electric Hybrid Ventilators, Wind Power, InTech, ISBN: 978-953-7619-81-7.

2 A. Aldawoud, R. Clark, 2008 Comparative analysis of energy performance between courtyard and atrium in buildings, Energy and Buildings, 40 209 214 .

3 O. S. Asfour, M. B. Gadi, 2007 A comparison between CFD and network models for predicting wind-driven ventilation in buildings, Building and Environment, 42 4079 4085

4 A. A. Badran, 2003 Performance of cool towers under various climates in Jordan, Energy & Buildings, 35 1031 1035 .

5 M. N. Bahadori, 1994 Viability of wind towers in achieving summer comfort in the hot arid regions of the Middle East, Renewable Energy, 5 879 892 .

6 N. K. Bansal, R. Mathur, M. S. Bhandari, 1994 A study of solar chimney assisted wind tower systems for natural ventilation in buildings, Building & Environment, 29 495 500 .

7 V. Baskaran, P. Bradshaw, 1993 Effect of bluff body wakes on skin friction and surface heat transfer in a turbulent boundary layer. Journal of Wind Engineering and Industrial Aerodynamics, 49 269 278

8 N. S. Billington, 1982 Energy efficient domestic ventilation systems for achieving acceptable indoor air quality, 3rd AIC Conference, September 20 23 , London, UK

9 B. O. Brooks, W. F. Davis, 1992 Understanding indoor air quality, CRC Press Inc., Boca Raton, Fla.

10 A. A. Elmualim, 2006 Dynamic modeling of a wind catcher/tower turret for natural ventilation, Building Serv. Eng. Res. Technol., 27 165 182 .

11 C. Faye, et al. 2005 Spitler, Heating, Ventilating and Air Conditioning: analysis and design (6th edition), John Wiley and Sons, Hoboken, NJ.

12 T. G. Flynn, N. A. Ahmed, 2005 Investigation of Rotating Ventilator using Smoke Flow Visualisation and Hot-wire anemometer. Proc. of 5th Pacific Symposium on Flow Visualisation and Image Processing Whitsundays, Australia, Paper PSFVIP-5 5 214 .

13 R. Greeno, 1997 Building Services, Technology and Design, Longman Publishing, Edinburgh

14 S. O. Hanssen, 1997 Economical consequences of poor indoor air quality and its relation to the total building operational costs, EuroFMlIFMA Conference & Exhibition, Turin, Italy, 1.- 3.

15 P. Heiselberg, E. Bjorn, P. V. Nielsen, 2001 Impact of open windows on room airflow and thermal comfort, International Journal of Ventilation, 1 91 100 .

16 P. Karave, T. Stathopoulos, A. K. Athienitis, 2005 Wind induced natural ventilation analysis, Solar Energy, 81 20 30 .

17 N. Khan, Y. Su, S. B. Riffat, C. Biggs, 2008 Performance testing and comparison of turbine ventilators. Renewable Energy, 33 2441 2447 .

18 M. Kolokotroni, A. Ayimomitis, Y. T. Ge, 2002 The suitability of wind driven natural ventilation towers for modern offices in the UK: a case study, World Renewable Energy Congress, 8

19 C. Lai, 2003 Experiments on the ventilation efficiency of turbine ventilator used for building and factory ventilation. Energy Buildings, 35 927 932 .

20 C. M. Lai, 2006 Prototype development of the rooftop turbine ventilator powered by hybrid wind and photovoltaic. Energy & Buildings, 38 174 180 .

21 L. Li, C. M. Mak, 2005 The assessment of the performance of a wind catcher system using CFD, Building & Environment, 42 1135 1141.

22 S. J. Lien, N. A. Ahmed, 2003 Application of multi-hole pressure probes in skin friction determination. AIAA conference paper; San Francisco: AIAA-2006 3659

23 J. Y. Luo, R. I. Issa, A. D. Gosman, 1994 Prediction of impeller-induced flows in mixing vessels using multiple frames of reference. I Chem E Symposium Series, 136 549 556 .

24 A. Mochida, et al. 2005 Methods for controlling airflow in and around a building under cross ventilation to improve indoor thermal comfort, Journal of Wind Engineering and Industrial Aerodynamics, 93 437 449

25 E. Olsen, Q. Chen, 2003 Energy consumption and comfort analysis for different low energy cooling systems in a mild climate, Energy & Buildings, 35 561 571 .

26 Larsen. G. Pillgram, 1991 300000 pa skolebenken for a bedre innemijoet, BE-nytt nr 2, Statens Bygningstekniske Etat, Oslo(In Norwegian)

27 A. Pisasale, N. A. Ahmed, 2002a Theoretical calibration for highly three-dimensional low-speed flows of a five-hole probe, Meas. Sci. Technol, 13 1100 1107 .

28 A. Pisasale, N. A. Ahmed, 2002b A novel method of extending the calibration range of five hole probe for highly three dimensional flows, J. of Flow Measurement and Instrumentation, 13 1), 23 30 .

29 A. Pisasale, N. A. Ahmed, 2003 Examining the effect of flow reversal on seven-hole probe measurements, AIAA Journal, 41 12), 2460 2467 .

30 A. Pisasale, N. A. Ahmed, 2004 Development of a functional relationship between port pressures and flow properties for the calibration and application of multihole probes to highly three-dimensional flows, Experiments of Fluids, 36 422 436 .

31 I. Rajapaksha, H. Nagai, M. Okumiya, 2003 A ventilated courtyard as a passive cooling strategy in the warm humid tropics, Renewable Energy, 28 1755 1778 .

32 D. M. H. Rashid, N. A. Ahmed, 2003 Study of Aerodynamic forces on a rotating wind driven ventilator. Journal of wind engineering, 27 63 72 .

33 A. Revel, 1998 Testing of 2 wind driven roof ventilators, available from www.edmonds.com.au (accessed 13/03/2007).

34 O. Seppanen, J. Palonen, 1998 The effects of indoor climate on national economy in Finland in billoon FIM, FISIAW 1998 SIY Rapportti 10, 1237-1866 1237 1866

35 S. West, 2001 Improving the sustainable development of building stock by the implementation of energy efficient, climate control technologies, Building & Environment, 36 281 289 .

Index